# **FE** Electrical & Computer

## Full-length Practice Exam

*Additional practice available at PrepFE.com*

PrepFE

Myles Cupp, Jon Van Lew, Damien Sutevski and the PrepFE Team

**ISBN 978-1-7345555-1-6**

# Contents

# About PrepFE

## PrepFE.com

PrepFE™ is primarily an online FE exam preparation service with various interactive practice exams, analytics, and a team devoted to helping you pass your exam. We believe the best way to prepare for the FE exam is with repeated practice, solving the types of problems that are on the exam. This promotes active instead of passive studying. While there's value in viewing lectures and presentations, active problem solving develops your skills, strengthens your memory recall, and builds your confidence. This was fundamental in our motivation for creating PrepFE.com. And we encourage you to visit for additional exam preparation.

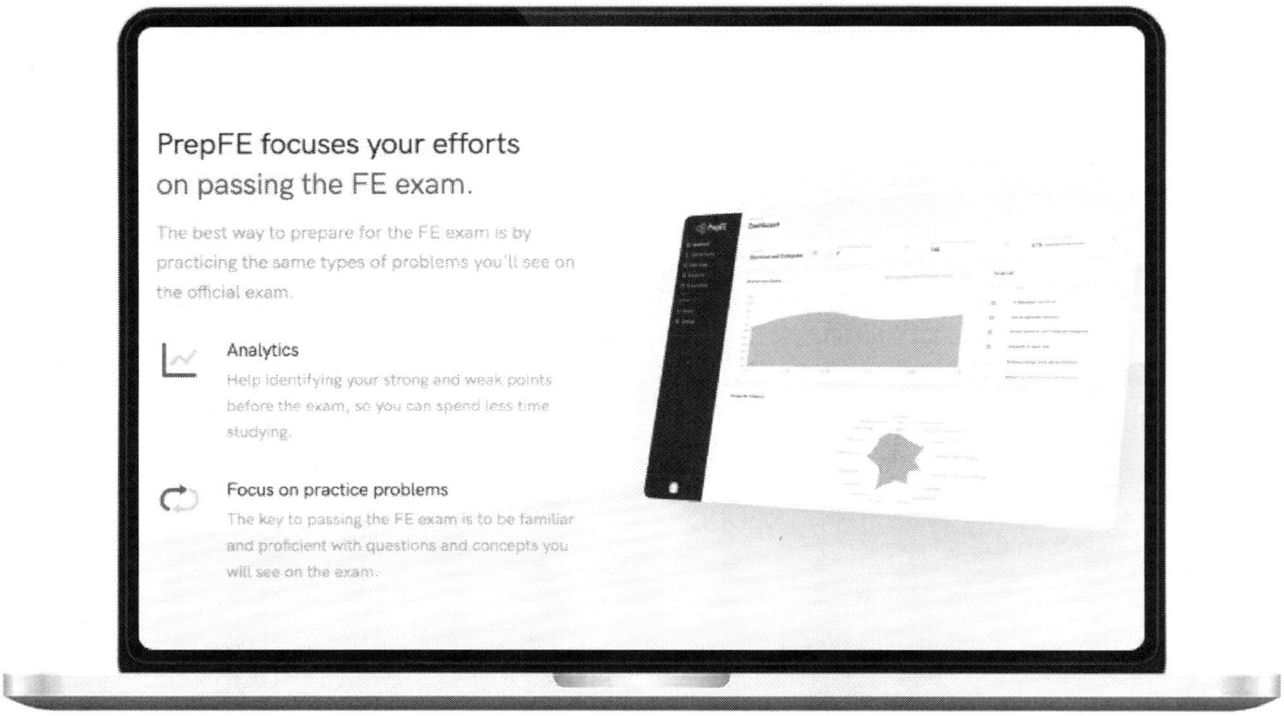


## PrepFE's Full-length Practice Exam Book

One useful and popular resource for preparing for the FE exam is the practice exam book provided by NCEES®, the administrators of the FE exam. At PrepFE, we wanted to create a book-based exam with the added benefit of performance analytics available on PrepFE.com. Problems come with a detailed solutions that explain how to solve them step-by-step, including shortcuts and tips for solving the problem easily and fast.

Prior to their publication here, all the problems in this book were benchmarked in order for us to refine them and gauge performance and difficulty. This enabled us to create a balanced practice exam that is most representative of the official FE exam – and comparing our benchmarks with publications from the NCEES, we think we've done just that!

We took book-based exams to the next level and built an online Performance Analysis feature where you can input your answers from this book and receive a visual report of how you stack up to others and to the NCEES's reported scores. You'll find more information about this in the Performance Analysis section (see page 7) or at PrepFE.com.

Good luck on your FE exam!

# Introduction

The official NCEES® Fundamentals of Engineering (FE) exam consists of 110 problems, and you are given 5 hours and 20 minutes, with an allotted break time, to work through all the problems. During the exam, you may flag problems and return to them later. However, the FE exam is broken into two halves. When you finish the first half and begin the second, you may not return to answer or change problems from the first half. For more information regarding the FE exam, please visit PrepFE.com and the NCEES's site, ncees.org.

For this PrepFE™ practice exam book, we recommend that you apply the same rule set you would experience in the official exam. Find a 6-hour window during which you can take the full exam in one sitting. If you get stuck on a problem, flag it and move on. Focus on topics and problems you are most comfortable with, and solve them first. Then go back and work on problems you flagged. But once you pass the half-way point, do not go back to change anything from the first half (we'll let you know when you've reached the halfway point).

When you take your official FE exam, make sure to answer every problem! There is no penalty for incorrect answers. Eliminating any seemingly-incorrect answers will increase your chances of guessing the right one. For this practice exam book, however, if you can't eliminate any wrong answers, and it's a completely random guess, we recommend that you do not choose any answer. You'll get a better self-assessment this way, especially if you use the Performance Analysis feature at PrepFE.com described next.

Have your *FE Reference Handbook* available and an NCEES-approved calculator on-hand.

## Performance Analytics

The NCEES does not provide official scores required to pass. However, at PrepFE.com, you may submit your final answers from this book, and we'll provide you with a personalized Performance Analysis. This analysis takes into account our benchmarked scores, others' submitted scores, and data provided from the NCEES – all broken down by topic, revealing your strengths and weaknesses. You'll also be able to provide any feedback, thoughts, or questions to us on the website as well. After you finish this practice exam, either visit the link provided in this QR code or go to `https://prepfe.com/performance` and submit your answers for your own performance review!

Here's an example of the results provided on PrepFE.com,

| Category | # of Problems | Score | Performance (0-15) | Average | Average Compared to FE examinees |
|---|---|---|---|---|---|
| Mathematics | 11 | 7 | 9.1 | 10.1 | |
| Probability and Statistics | 4 | 4 | 15.0 | 10.3 | |
| Ethics and Professional Practice | 4 | 4 | 15.0 | 11.8 | |
| Engineering Economics | 5 | 5 | 15.0 | 9.9 | |
| Properties of Electrical Materials | 4 | 2 | 6.7 | 11.0 | |
| Circuit Analysis (DC and AC Steady State) | 12 | 9 | 11.2 | 9.9 | |
| Linear Systems | 5 | 3 | 8.7 | 9.4 | |
| Signal Processing | 6 | 3 | 7.3 | 9.7 | |
| Electronics | 8 | 3 | 5.1 | 9.7 | |
| Power Systems | 10 | 7 | 9.8 | 9.6 | |
| Electromagnetics | 4 | 1 | 3.8 | 10.0 | |
| Control Systems | 6 | 5 | 12.4 | 9.2 | |
| Communications | 7 | 3 | 6.6 | 9.0 | |
| Computer Networks | 6 | 5 | 12.2 | 9.5 | |
| Digital Systems | 9 | 8 | 13.1 | 9.2 | |
| Computer Systems | 5 | 2 | 6.1 | 8.9 | |
| Software Engineering | 4 | 3 | 11.7 | 10.5 | |
| | | 74 | 9.9 | 9.8 | |

See your own results at `prepfe.com/performance`

As a convenience for you, on the last page of this book, we provide you with a blank answer sheet that you can fill out while taking the exam. You may use this answer sheet as a reference when submitting your work to PrepFE.com.

## Errata

We've worked rigorously to create a full-length exam that is representative of the official FE exam, with clear problem prompts and solutions, and without errors. We realize that some mistakes or oversights may have made it through our due diligence, so an online errata webpage will be maintained for any issues found or reported.

For any errors found, clarifications, or other changes to the problems in this exam book, please visit `https://prepfe.com/errata`, or follow the QR code given here.

## Additional Practice

At PrepFE, we strongly encourage preparing for the FE exam by practicing as many problems as you can. Active and engaged learning like this is the best way to prepare for the FE exam. We provide hundreds of problems with detailed and thorough solutions and explanations, and you can practice with several exam variations, such as timed exams or with a focus on specific categories.

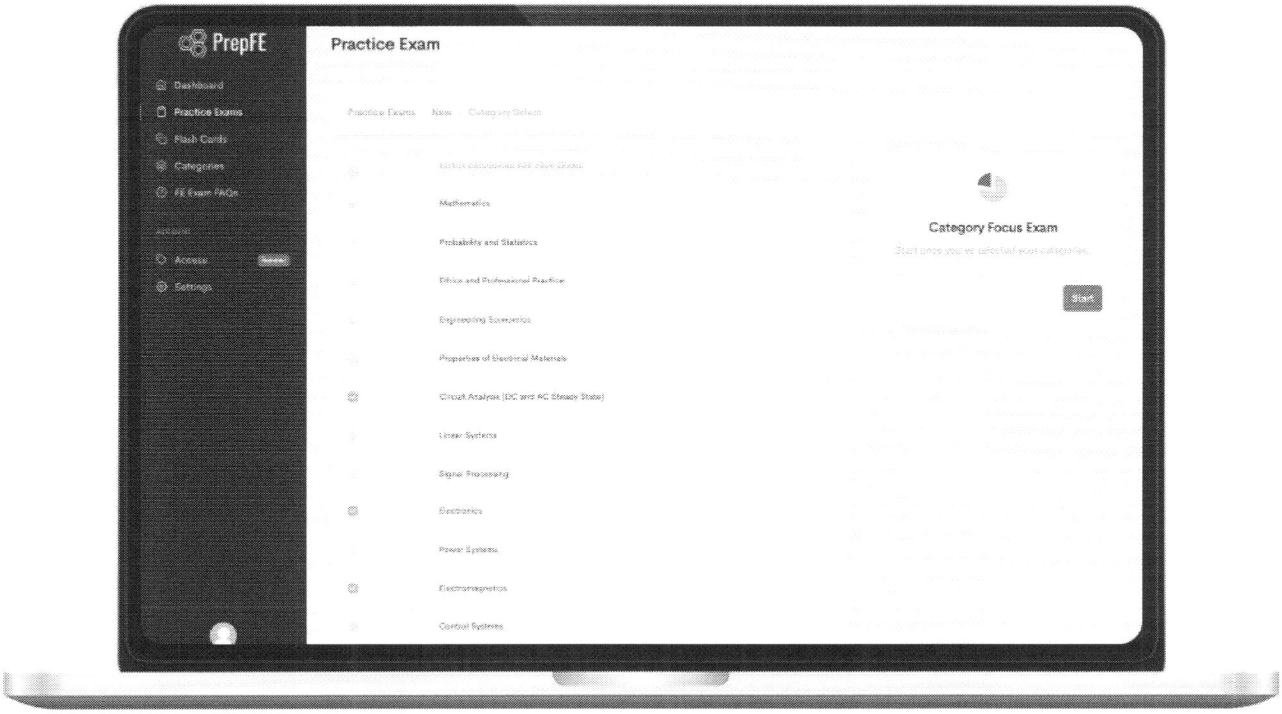


Visit PrepFE.com to learn more!

# Topics Overview

| Topic | Number of questions |
|---|---|

1. **Mathematics** **11-17**

   Algebra and trigonometry
   Complex numbers
   Discrete mathematics
   Analytic geometry
   Calculus (*e.g.*, differential, integral, single-variable, multivariable)
   Ordinary differential equations
   Linear algebra
   Vector analysis

2. **Probability and Statistics** **4-6**

   Measures of central tendencies and dispersions (*e.g.*, mean, mode, standard deviation)
   Probability distributions (*e.g.*, discrete, continuous, normal, binomial, conditional probability)
   Expected value (weighted average)

3. **Ethics and Professional Practice** **4-6**

   Codes of ethics (*e.g.*, professional and technical societies, NCEES *Model Law* and *Model Rules*)
   Intellectual property (*e.g.*, copyright, trade secrets, patents, trademarks)
   Safety (*e.g.*, grounding, material safety data, PPE, radiation protection)

4. **Engineering Economics** **5-8**

   Time value of money (*e.g.*, present value, future value, annuities)
   Cost estimation
   Risk identification
   Analysis (*e.g.*, cost-benefit, trade-off, break-even)

5. **Properties of Electrical Materials** **4-6**

Semiconductor materials (*e.g.*, tunneling, diffusion/drift current, energy bands, doping bands, p-n theory)

Electrical (*e.g.*, conductivity, resistivity, permittivity, magnetic permeability, noise)

Thermal (*e.g.*, conductivity, expansion)

6. **Circuit Analysis (DC and AC Steady State)** **11-17**

KCL, KVL

Series/parallel equivalent circuits

Thevenin and Norton theorems

Node and loop analysis

Waveform analysis (*e.g.*, RMS, average, frequency, phase, wavelength)

Phasors

Impedance

7. **Linear Systems** **5-8**

Frequency/transient response

Resonance

Laplace transforms

Transfer functions

8. **Signal Processing** **5-8**

Sampling (*e.g.*, aliasing, Nyquist theorem)

Analog filters

Digital filters (*e.g.*, difference equations, Z-transforms)

9. **Electronics** **7-11**

Models, biasing, and performance of discrete devices (*e.g.*, diodes, transistors, thyristors)

Amplifiers (*e.g.*, single-stage/common emitter, differential, biasing)

Operational amplifiers (*e.g.*, ideal, nonideal)

Instrumentation (*e.g.*, measurements, data acquisition, transducers)

Power electronics (*e.g.*, rectifiers, inverters, converters)

10. **Power Systems** **8-12**

Power theory (*e.g.*, power factor, single- and three-phase, voltage regulation)

Transmission and distribution (*e.g.*, real and reactive losses, efficiency, voltage drop, delta and wye connections)

Transformers (*e.g.*, single-phase and three-phase connections, reflected impedance)

Motors and generators (*e.g.*, synchronous, induction, DC)

11. **Electromagnetics** **4-6**

Electrostatics/magnetostatics (*e.g.*, spatial relationships, vector analysis)
Electrodynamics (*e.g.*, Maxwell equations, wave propagation)
Transmission lines (high frequency)

12. **Control Systems** **6-9**

Block diagrams (*e.g.*, feedforward, feedback)
Bode plots
Closed-loop response, open-loop response, and stability
Controller performance (*e.g.*, steady-state errors, settling time, overshoot)

13. **Communications** **5-8**

Basic modulation/demodulation concepts (*e.g.*, AM, FM, PCM)
Fourier transforms/Fourier series
Multiplexing (*e.g.*, time division, frequency division, code division)
Digital communications

14. **Computer Networks** **4-6**

Routing and switching
Network topologies (*e.g.*, mesh, ring, star)
Network types (*e.g.*, LAN, WAN, internet)
Network models (*e.g.*, OSI, TCP/IP)
Network intrusion detection and prevention (*e.g.*, firewalls, endpoint detection, network detection)
Security (*e.g.*, port scanning, network vulnerability testing, web vulnerability testing, penetration testing, security triad)

15. **Digital Systems** **8-12**

Number systems
Boolean logic
Logic gates and circuits
Logic minimization (*e.g.*, SOP, POS, Karnaugh maps)
Flip-flops and counters
Programmable logic devices and gate arrays
State machine design
Timing (*e.g.*, diagrams, asynchronous inputs, race conditions and other hazards)

16. **Computer Systems** **5-8**

Microprocessors
Memory technology and systems
Interfacing

17. **Software Engineering** **4-6**

Algorithms (*e.g.*, sorting, searching, complexity, big-O)

Data structures (*e.g.*, lists, trees, vectors, structures, arrays)

Software implementation (*e.g.*, iteration, conditionals, recursion, control flow, scripting, testing)

# Practice Exam

Remember that this practice exam is your chance to run through the FE exam under similar conditions to the real deal. We suggest that you set aside 6 hours and run through this entire exam in one sitting, keep track of your time, and adhere to the rule about not revising the first-half problems once you begin the second half (we'll notify you at the half-way point). Lastly, there is a blank answer sheet at the very back of this book you may use to record your answers as you go.

Take a deep breath, get comfortable, and good luck!

**1.** The possible value(s) for $x$ in the polynomial shown below is/are most nearly:

$$3x = 10 - x^2$$

- A. 4.3, 5
- B. -2, -5
- C. 2, -5
- D. No domain

**2.** The rectangular coordinates of a complex number are $(6, 9)$. What are the complex number's approximate polar coordinates?

- A. $0.33\angle 45.21°$
- B. $16.82\angle 36.31°$
- C. $6\angle 30°$
- D. $10.82\angle 56.31°$

**3.** Given the sets $A = \{1, 3\}$ and $B = \{2, 4\}$, find the Cartesian product $A \times B$.

- A. $\{(2, 12)\}$
- B. $\{(1, 2)(1, 4)(3, 2)(3, 4)\}$
- C. $\{(1, 3)(2, 4)\}$
- D. $\{(2, 1)(2, 3)(4, 1)(4, 3)\}$

**4.** If $A = \{1, 3, 5, 7\}$ and $B = \{1, 5, 9, 10\}$ then $A \cap B$ is:

- A. $\{1, 3, 5, 7, 9, 10\}$
- B. $\{3, 9, 10\}$
- C. $\{1, 5\}$
- D. None of these

**5.** The traverse below illustrates a property line's coordinates. The distance between points A and C is most nearly ________

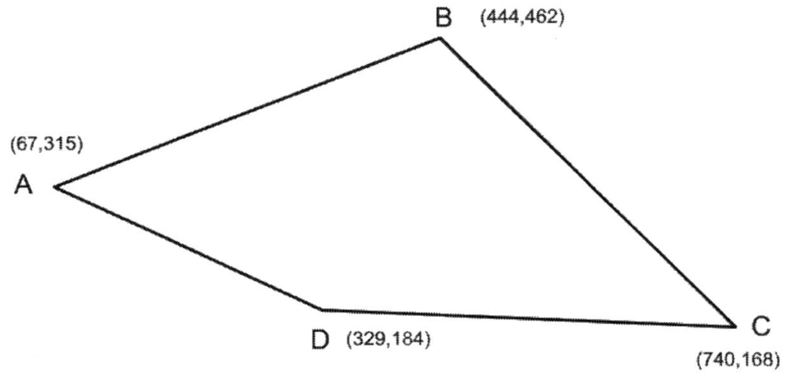


Enter your answer on the blank, rounded to the nearest integer if necessary.

**6.** What is the area between the line $x_0 = 1.5$, the x- and y-axes, and the curve $f(x) = -x^3$ shown?

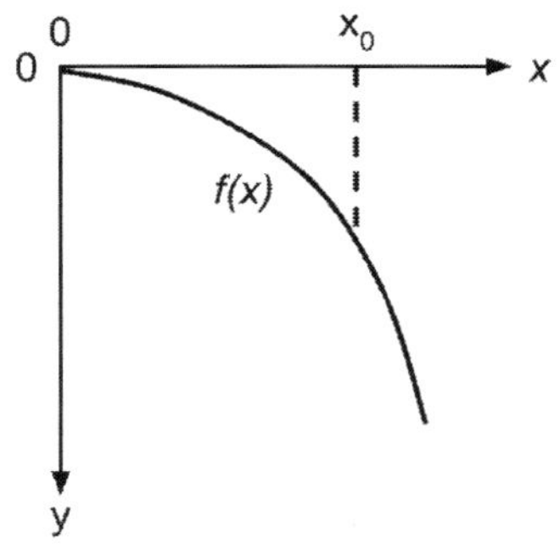


- A. -1.27
- B. -.37
- C. -.76
- D. -1.05

**7.** What is most nearly the derivative of $\tan(5x)$?

- A. $5\sec^2(5x)\tan(x)$
- B. $5\csc^2(5x)$
- C. $\sec^2(5x)$
- D. $5\sec^2(5x)$

**8.** Provide the solution for the ordinary differential equation given as:

$$\frac{dx}{dt} = 5x - 3$$

- A. $x(t) = Ce^{5t} + \frac{3}{5}$
- B. $x(t) = \frac{1}{5}Ce^{\frac{5}{3}t}$
- C. $x(t) = \frac{1}{5}\log|5t - 3| + C$
- D. $x(t) = Ce^{5t}$

**9.** Determine the general solution to the given differential equation:

$$y^2 + x^2y' = 0$$

- A. $x + y = C$
- B. $\frac{1}{x^2} + \frac{1}{y^2} = C$
- C. $\frac{1}{x} + \frac{1}{y} = C$
- D. $\frac{1}{x} + \frac{1}{y^3} = C$

**10.** The following matrix has a rank of ________.

$$A = \begin{bmatrix} 3 & -4 & -6 \\ 2 & 4 & -4 \\ 4 & 6 & -8 \end{bmatrix}$$

Enter your answer on the blank

**11.** A vector field **D** given as:

$$\mathbf{D} = 5xy\mathbf{i} + 2xy^3z\mathbf{j} + x^3z\mathbf{k}$$

The divergence of this field is most nearly:

- A. $5xy + 2xy^3z + x^3z$
- B. $5 + 6y^2 + 3x^2$
- C. $5x + 6x^2 + 3x^2$
- D. $x^3 + 6xy^2z + 5y$

**12.** Given the following dataset,

$$0.61, 0.74, 0.76, 0.55, 0.80, 0.55, 0.60$$

The mode of the dataset is ________.

The median of the dataset is ________.

The mean of the dataset is ________.

Enter your answers on the blanks, provide your answer to the hundredths place.

**13.** What is the probability of heads being flipped *at least once* after a coin is tossed three times?

- ○ *A.* 67%
- ○ *B.* 22%
- ○ *C.* 50%
- ○ *D.* 88%

**14.** Alejandro is asked to determine the retention rate for an app's users given the following two metrics:

1. Anyone who downloads the app is 95% likely to use the app for 1 week.
2. Anyone who downloads the app is 80% likely to use the app for 1 month.

What is the probability that someone uses the app for 1 month given they've tried the app for 1 week?

- ○ *A.* 91%
- ○ *B.* 90%
- ○ *C.* 84%
- ○ *D.* 80%

**15.** A recent 1,200-person poll showed 22% of interviewed college students approve having 8AM classes. Two polled college students were later selected at random. The discrete Bernoulli probability distribution whether 0, 1, or 2 of the 2 interviewed students approves having 8AM classes is as follows:

| # of students out of 2 | 0 students approving | 1 student approving | 2 students approving |
|---|---|---|---|
| P(x) | 60.84% | 34.20% | 4.84% |

The variance of the expected mean value is most nearly:

- ○ *A.* 0.34
- ○ *B.* 0.44
- ○ *C.* 1.2
- ○ *D.* 1.5

**16.** Coverage for bodily injury and property damage claims made by others describes:

- ◯ *A.* a bond
- ◯ *B.* indemnity
- ◯ *C.* a lien
- ◯ *D.* liability insurance

**17.** You, a licensed professional engineer (PE), have signed and stamped a few engineering documents for projects in Utah. The existing client now has few more projects to be built but now in Florida. Under which conditions would you, a licensed professional engineer (PE), be allowed to sign/seal engineering documents created in Utah but to be built in Florida?

- ◯ *A.* No new conditions since the engineering design is happening from Utah, where the PE is licensed
- ◯ *B.* No new conditions since client and PE have worked together in the past
- ◯ *C.* You relocate to Florida
- ◯ *D.* You apply for PE licensure in Florida

**18.** An engineer attends a trade convention and demonstrates a new invention to potential investors. Is their invention still patentable?

- ◯ *A.* Yes, if they file the patent within 12 months of the public demonstration
- ◯ *B.* Yes, if they file the patent within 24 months of the public demonstration
- ◯ *C.* No, they have made a public disclosure; the invention is no longer patentable
- ◯ *D.* No, because the investors now claim ownership of the invention after attending the demonstration

**19.** Safety glasses are important to wear while working on electrical circuits for which of the following reasons?

- ◯ *A.* Protection from shrapnel in the event of an explosion
- ◯ *B.* Protection from sparks should a short-circuit occur
- ◯ *C.* Protection from exposure to chemical irritants
- ◯ *D.* All of these

**20.** A new company machine is purchased for $15,000. Using MACRS, about how much does it depreciate in 3 years if it has an expected lifetime of 10 years?

- ◯ *A.* $13,888
- ◯ *B.* $2,221
- ◯ *C.* $2,160
- ◯ *D.* $6,360

**21.** \$57,000 from an inheritance is deposited into a CD bank account that has an interest rate of 2.1%, compounded semi-annually. If no more deposits are ever made, the money earned from interest after 10 years is most nearly:

- ○ *A.* \$13,167
- ○ *B.* \$70,236
- ○ *C.* \$13,236
- ○ *D.* \$18,460

**22.** You recently purchased a home rental property for a lump sum of \$250,000. As it is, the home could be rented for \$2,000/month. However, you are considering if it's financially worth it to fix up the home, so it could be rented for more. If the home is remodeled, it could be rented for \$2,800/month. The cost to remodel the home is \$70,000. Neglecting a home resell value, which alternative would yield a higher rate-of-return interest on the invested amount after 15 years?

- ○ *A.* Remodeling the home and renting it for \$2,800/month
- ○ *B.* Renting the home as it is for \$2,000/month
- ○ *C.* Same rate of return between remodeled and non-remodeled home
- ○ *D.* Cannot determine rates of return with the information supplied

**23.** A mom n' pop flower nursery has an annual operation cost of \$200,000. It costs the flower nursery \$50 in material costs to put together one bouquet of flowers. They sell the flower bouquets for \$125 each. About how many bouquets do they need to sell to break even?

- ○ *A.* 2,667 units
- ○ *B.* 2,350 units
- ○ *C.* 1,910 units
- ○ *D.* 1,142 units

**24.** The life-cycle costing method includes the initial cost and what other cost value?

- ○ *A.* Green Construction Cost
- ○ *B.* Present value of periodic repair costs and operating costs
- ○ *C.* Carbon emission costing
- ○ *D.* Accelerated Construction Cost

**25.** A p-n junction at 300 K is doped on the n-side by a factor of 5000 and the p-side remains unchanged in doping. What is the resultant change in the barrier height of the p-n junction?

- ○ *A.* 5000 mV
- ○ *B.* 500 mV
- ○ *C.* 109 mV
- ○ *D.* 221 mV

**26.** At 90°F, the contact potential for a given p-n junction is 0.030 V. If the temperature is raised to 190°F, the increase in mV in the contact potential is most nearly:

- ○ *A.* 1.344 mV
- ○ *B.* 2.468 mV
- ○ *C.* 30.01 mV
- ○ *D.* 5.458 mV

**27.** The permittivity of a material is measured to be $10.12 \times 10^{-12}\,\mathrm{F\,m^{-1}}$. The value of $\kappa$ is most nearly:

- ○ *A.* 1.14
- ○ *B.* 0.87
- ○ *C.* 1.02
- ○ *D.* 0.67

**28.** Calculate the temperature coefficient of a given material when its resistance triples with a 30 K temperature increase:

- ○ *A.* $0.013\,\mathrm{K^{-1}}$
- ○ *B.* $0.222\,\mathrm{K^{-1}}$
- ○ *C.* $0.066\,\mathrm{K^{-1}}$
- ○ *D.* $0.120\,\mathrm{K^{-1}}$

**29.** A series loop contains the following circuit elements in order: a 5 V source, a 3 kΩ resistor, a 5 kΩ resistor, a 20 V source, and a 7 kΩ resistor. The magnitude of the voltage across each resistor is most nearly:

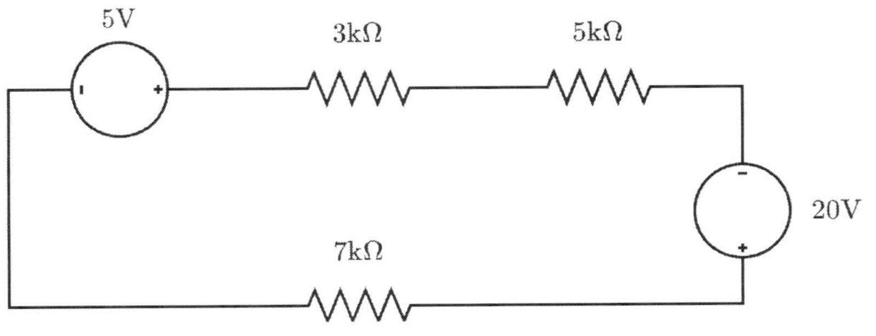


- ○ *A.* $v_{3\,\mathrm{k\Omega}} = 1\,\mathrm{V}, v_{5\,\mathrm{k\Omega}} = 7\,\mathrm{V}, v_{7\,\mathrm{k\Omega}} = 9\,\mathrm{V}$
- ○ *B.* $v_{3\,\mathrm{k\Omega}} = 5\,\mathrm{V}, v_{5\,\mathrm{k\Omega}} = 20\,\mathrm{V}, v_{7\,\mathrm{k\Omega}} = 0\,\mathrm{V}$
- ○ *C.* $v_{3\,\mathrm{k\Omega}} = 2\,\mathrm{V}, v_{5\,\mathrm{k\Omega}} = 3.125\,\mathrm{V}, v_{7\,\mathrm{k\Omega}} = 5.325\,\mathrm{V}$
- ○ *D.* $v_{3\,\mathrm{k\Omega}} = 5\,\mathrm{V}, v_{5\,\mathrm{k\Omega}} = 8.335\,\mathrm{V}, v_{7\,\mathrm{k\Omega}} = 11.669\,\mathrm{V}$

**30.** Determine the voltage across the $5\,\text{k}\Omega$ resistor in the circuit.

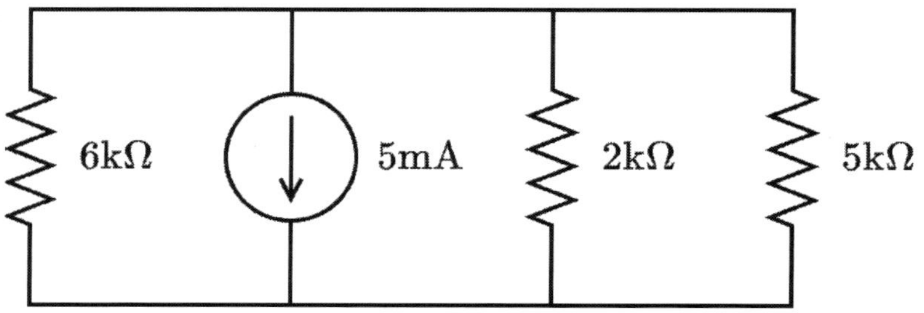


- A. $-4.78\,\text{V}$
- B. $3.78\,\text{V}$
- C. $-5.77\,\text{V}$
- D. $10.83\,\text{V}$

**31.** Find the equivalent resistance R and the equivalent capacitance C for the given network:

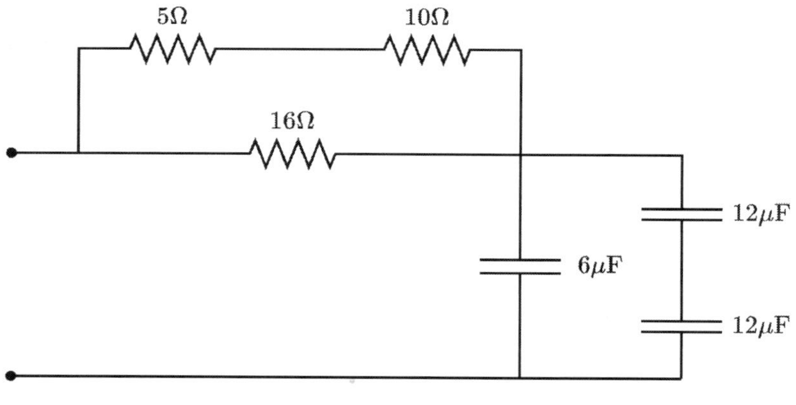


- A. $R_{\text{eq}} = 8.0\,\Omega, C_{\text{eq}} = 16\,\mu\text{F}$
- B. $R_{\text{eq}} = 7.7\,\Omega, C_{\text{eq}} = 6\,\mu\text{F}$
- C. $R_{\text{eq}} = 31\,\Omega, C_{\text{eq}} = 30\,\mu\text{F}$
- D. $R_{\text{eq}} = 7.7\,\Omega, C_{\text{eq}} = 12\,\mu\text{F}$

**32.** The equivalent resistance, $R_{eq}$, at terminals a-b is most nearly:

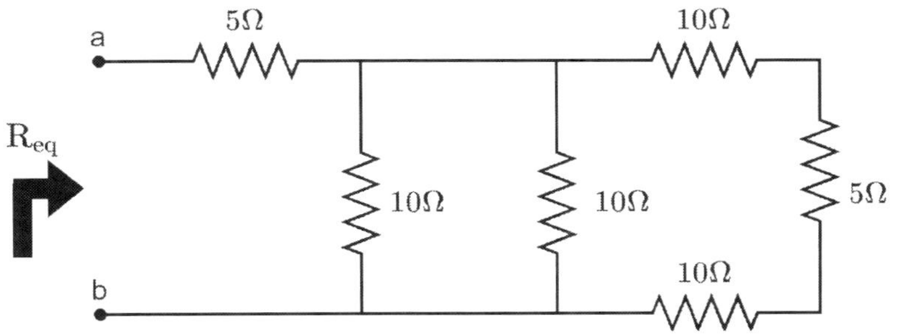


- A. 4.17 Ω
- B. 9.17 Ω
- C. 15.0 Ω
- D. 7.14 Ω

**33.** The Thevenin equivalent resistance Ω at Points A-B is most nearly ________.

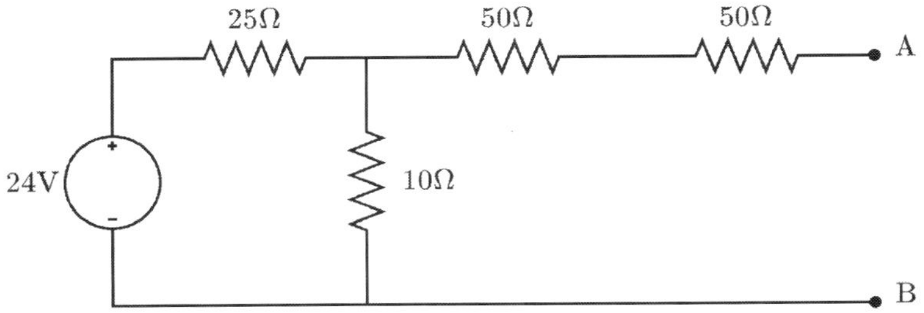


Enter your answer on the blank, rounded to the nearest integer if necessary.

**34.** Determine the voltage at Node A.

$$R_1 = 80\,\Omega\ ;\ R_2 = 60\,\Omega\ ;\ R_3 = 40\,\Omega\ ;\ I = 2\,\text{A}$$

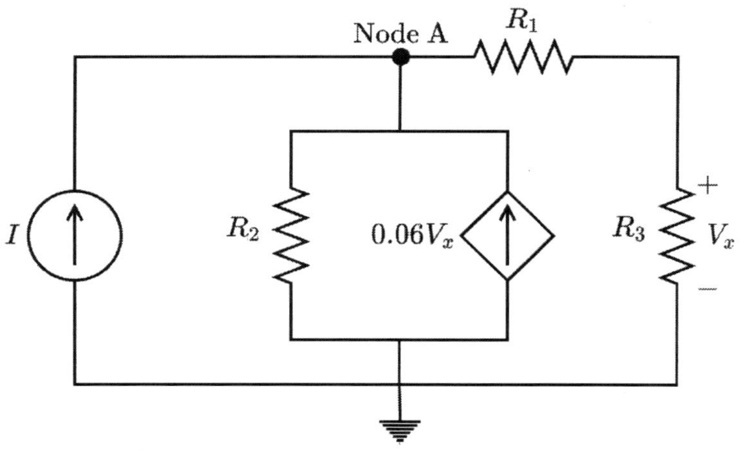


- A. 400 V
- B. 300 V
- C. 200 V
- D. 100 V

**35.** Consider the following circuit in the time-domain.

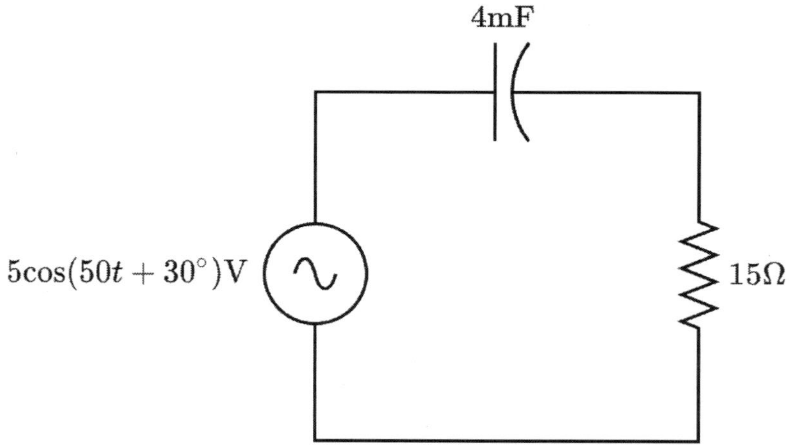


The component values in the frequency domain are most nearly:

- A. $Z_C = -j5\,\Omega;\ Z_R = 15\,\Omega;\ V_S = 5\angle 30°\text{V}$
- B. $Z_C = -j0.2\,\Omega;\ Z_R = 15\,\Omega;\ V_S = 5\angle 30°\text{V}$
- C. $Z_C = j5\,\Omega;\ Z_R = 15\,\Omega;\ V_S = 30\angle 50°\text{V}$
- D. $Z_C = -j5\,\Omega;\ Z_R = 15\,\Omega;\ V_S = 3.53\angle 30°\text{V}$

**36.** The effective (rms) value of the periodic current waveform is most nearly:

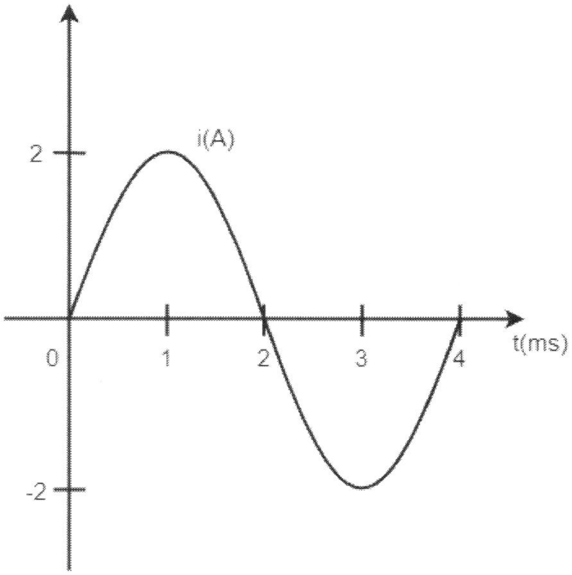


- A. 1.41 A
- B. 1.83 A
- C. 4.00 A
- D. 2.00 A

**37.** Find the voltage source necessary to produce the response in the given circuit:

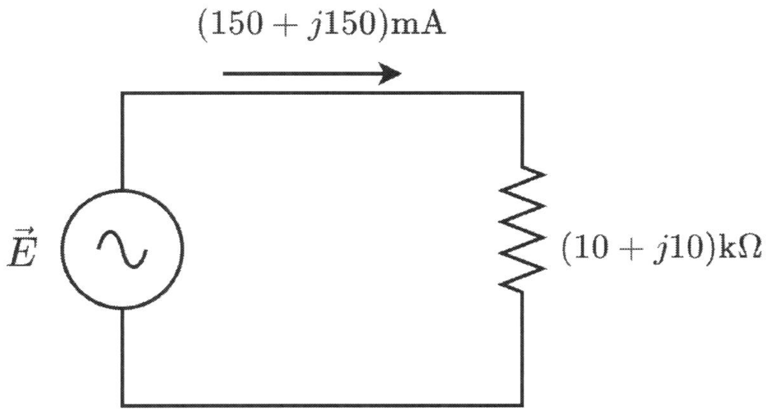


- A. $900\angle 60°$V
- B. $1500\angle 90°$V
- C. $3800\angle 0°$V
- D. $3000\angle 90°$V

**38.** In a phasor diagram, the length of the phasor generally represents:

- ○ *A.* the magnitude of the maximum value
- ○ *B.* the magnitude of the maximum frequency
- ○ *C.* the magnitude of the phase angle
- ○ *D.* the magnitude of the rms value

**39.** Find the series impedance at $f = 120\,\text{kHz}$.

$$C = 120\,\text{pF}\ ;\ L = 15\,\text{mH}\ ;\ R = 500\,\Omega$$

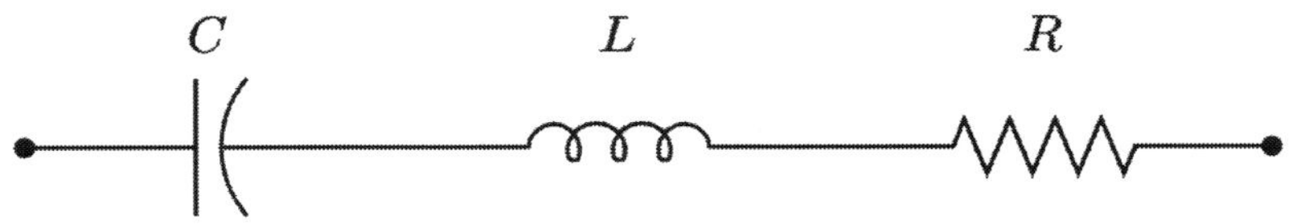


- ○ *A.* $563\angle 27.3°\Omega$
- ○ *B.* $317\angle 32.1°\Omega$
- ○ *C.* $620\angle 47.9°\Omega$
- ○ *D.* $520\angle 17.2°\Omega$

**40.** At 2000 rad/sec, then the admittance of a series combination of a $5\,\Omega$ resistor and a $100\,\mu\text{F}$ capacitor is most nearly:

- ○ *A.* $1.0 + j1.0\,℧$
- ○ *B.* $0.8 + j0.8\,℧$
- ○ *C.* $0.1 + j0.1\,℧$
- ○ *D.* $0.5 + j0.5\,℧$

**41.** An exponentially decaying voltage has a time constant of 400 ms and an initial value of 1.2 kV. What is the value of the voltage at $t = 125\,\text{ms}$?

- ○ *A.* 877.9 V
- ○ *B.* 532.4 V
- ○ *C.* 1135.1 V
- ○ *D.* 420.4 V

**42.** A coil with inductance and resistance of 1 mH and $2\,\Omega$, respectively, is connected in series to a capacitor and a 120 V, 5 kHz supply. The value of capacitance that will cause the series circuit to be in resonance is most nearly:

- ○ *A.* 3 µF
- ○ *B.* 2 µF
- ○ *C.* 1 µF
- ○ *D.* 0.5 µF

**43.** Compute the Laplace Transform of:

$$f(t) = u(t) - u(t-1)$$

- A. $1 + e^{-s}$
- B. $\frac{1}{s}[1 - e^{-s}]$
- C. $1 - e^{-s}$
- D. $\frac{1}{s^2}[1 - e^{-s}]$

**44.** Compute the Laplace Transform of:

$$f(t) = te^{-t}u(t)$$

- A. $\dfrac{1}{(s+1)^2}$
- B. $\dfrac{1}{(s+1)}$
- C. $\dfrac{1}{(s+2)^2}$
- D. $\dfrac{1}{(s-1)^2}$

**45.** Determine the transfer function T(s) of the following circuit if $R = 250\,\text{k}\Omega$ and $C = 6\,\mu\text{F}$:

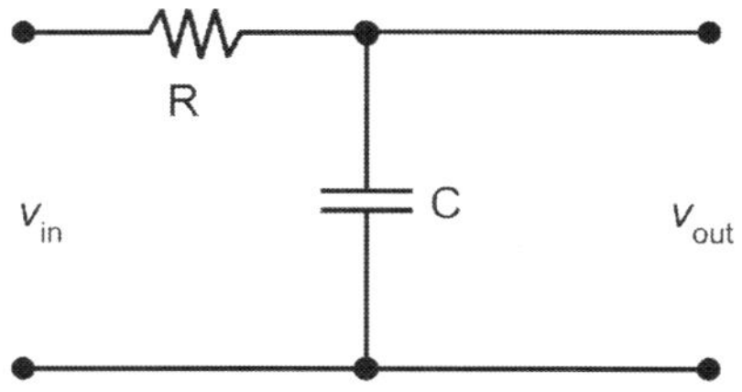


- A. $\dfrac{4}{3s+6}$
- B. $\dfrac{2}{3s+2}$
- C. $\dfrac{2}{3s+1}$
- D. $\dfrac{4}{3s+2}$

**46.** A sinusoidal signal in the frequency range of 0.5 to 5 kHz is being sampled at a rate equal to four times the Nyquist rate. What is the time between samples?

- ○ *A.* 2.5 µs
- ○ *B.* 20 µs
- ○ *C.* 25 µs
- ○ *D.* 10 µs

**47.** A compact disc (CD) records audio signals directly by using PCM. Assume the audio signal bandwidth is 20 kHz. What is the threshold for the Nyquist sampling rate?

- ○ *A.* 40 kHz
- ○ *B.* 20 kHz
- ○ *C.* 10 kHz
- ○ *D.* 30 kHz

**48.** What is most nearly the cutoff frequency of the given circuit?

$$R_1 = 10\,\text{k}\Omega$$
$$C_1 = 270\,\text{nF}$$

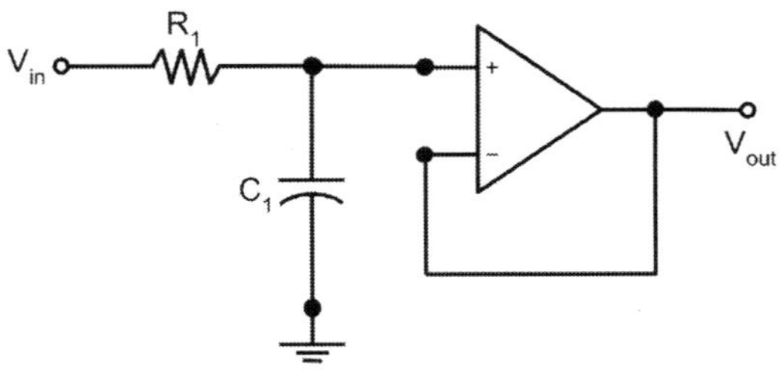


- ○ *A.* 90 Hz
- ○ *B.* 60 Hz
- ○ *C.* 30 Hz
- ○ *D.* 10 Hz

**49.** The transfer function for a filter is:

$$H(j\omega) = \frac{j\omega + 0.3}{j\omega + 5}$$

This filter is best classified as a:

- ○ *A.* high-Pass filter
- ○ *B.* low-Pass filter
- ○ *C.* band-Pass filter
- ○ *D.* notch filter

**50.** The difference equation of a digital filter is given by:

$$y_k = 2y_{k-1} - 3y_{k-2} + 2x_{k-1} + 0.05x_{k-2}$$

Determine the transfer function, $H(z)$.

- ○ *A.* $H(z) = \dfrac{1}{z^1 - z}$
- ○ *B.* $H(z) = \dfrac{2z + 0.05}{z^2 - 2z + 3}$
- ○ *C.* $H(z) = \dfrac{3z - z}{z - 2.05}$
- ○ *D.* $H(z) = \dfrac{2z^2 - 3z}{z^2 - 2z + 0.05}$

**51.** The first three terms for the given difference equation are:

$$y[k+1] + 2y[k] = f[k+1]$$

Assume $f[k] = e^{-k}u[k]$ and $y[-1] = 0$.

- ○ *A.* $1, -1.632, 3.399$
- ○ *B.* $2, 7.221, 1.462$
- ○ *C.* $0, 2.361, 4.210$
- ○ *D.* $1, 3.555, 2.109$

**52.** An n-channel JFET has a pinch-off voltage, $V_p = -2\,\text{V}$, and $I_{DSS} = 12\,\text{mA}$. This device is intended to operate in the saturation region. When the quiescent drain current is $2\,\text{mA}$, the gate-to-source voltage ($V_{GS}$) required to give this operating point is most nearly:

- ○ *A.* $-2.33\,\text{V}$
- ○ *B.* $-2.82\,\text{V}$
- ○ *C.* $-2.00\,\text{V}$
- ○ *D.* $-1.18\,\text{V}$

**53.** A diode is measured at two operating currents, 0.2 mA and 4 mA is found to have corresponding voltages of 0.70 V and 0.85 V at 300 K. The values of $n$ and $I_s$ are most nearly:

- ○ A. $n = 2.3$ ; $I_s = 150 \times 10^{-12}$ A
- ○ B. $n = 1.5$ ; $I_s = 120 \times 10^{-12}$ A
- ○ C. $n = 2.0$ ; $I_s = 170 \times 10^{-12}$ A
- ○ D. $n = 2.5$ ; $I_s = 150 \times 10^{-12}$ A

**54.** What is most nearly the equivalent AC collector resistance in the given circuit?
$R_1 = 2\,\text{k}\Omega$, $R_2 = 500\,\Omega$
$R_C = 700\,\Omega$, $R_E = 220\,\Omega$
$V_S = 2\,\text{mV}$, $R_S = 25\,\Omega$
$V_{CC} = 15\,\text{V}$, $R_L = 3\,\text{k}\Omega$

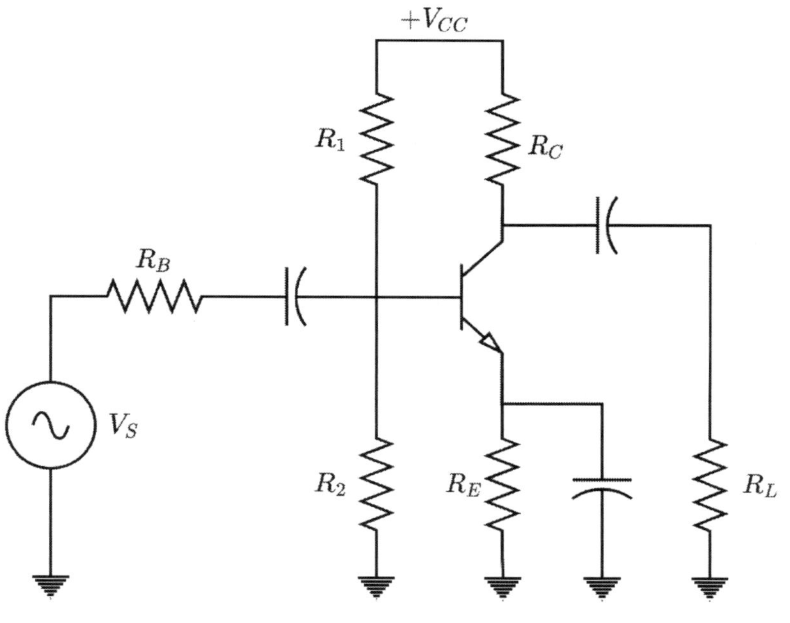


- ○ A. 700 Ω
- ○ B. 920 Ω
- ○ C. 1 kΩ
- ○ D. 568 Ω

**55.** In the circuit shown, $R_B = 5\,\mathrm{k\Omega}$, $V_B = 4\,\mathrm{V}$, $V_{BE} = 0.7\,\mathrm{V}$, $V_{CC} = 40\,\mathrm{V}$, and $R_E = 2\,\mathrm{k\Omega}$. For $\beta = 100$ and $R_C = 5\,\mathrm{k\Omega}$, determine the Q point for this transistor in the active region:

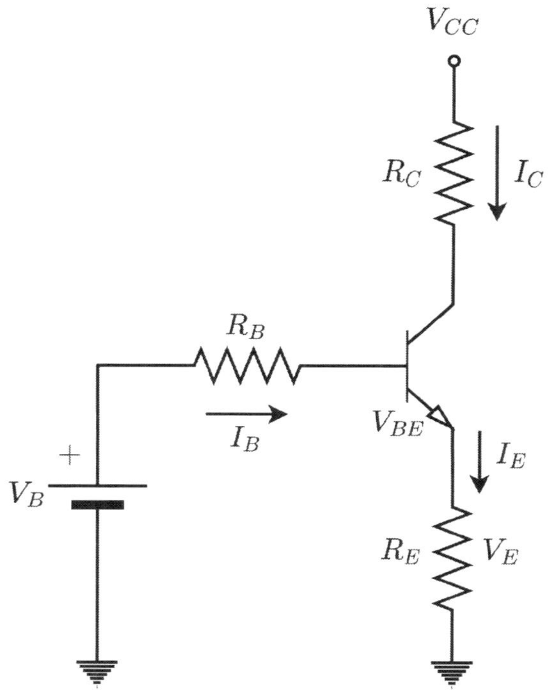


- A. $I_C = 1\,\mathrm{mA}, V_{CE} = 10\,\mathrm{V}$
- B. $I_C = 1.59\,\mathrm{mA}, V_{CE} = 28.8\,\mathrm{V}$
- C. $I_C = 5\,\mathrm{mA}, V_{CE} = 0\,\mathrm{V}$
- D. $I_C = 2.14\,\mathrm{mA}, V_{CE} = 14.6\,\mathrm{V}$

**56.** In an operational amplifier, the gain is given by $A = \frac{1000}{1+j\omega}$. What is the maximum value of the magnitude of $A$ and the frequency it occurs?

- A. $1000, f = 0$
- B. $\dfrac{1000}{1+j}, f = 100\,\mathrm{rad\,s^{-1}}$
- C. $0.001, f = 0$
- D. $\dfrac{1000}{1+j2\pi}, f = 1000\,\mathrm{rad\,s^{-1}}$

# END OF THE FIRST HALF

To get the most from this practice exam, we highly recommend you treat this as you would the real exam. Before you advance to the second half, take a moment, and make certain you are satisfied with your answers in the first half.

After you begin the second-half problems, don't go back and work on any problems before this point, even if you have extra time at the end of the exam. This will give you a fair evaluation of where you stand in your preparedness for the FE Exam.

**57.** A portion of an electronic sensing and amplifying circuit has the given setup. The expression that correctly represents the output, $v_{out}$, of the circuit is most nearly:

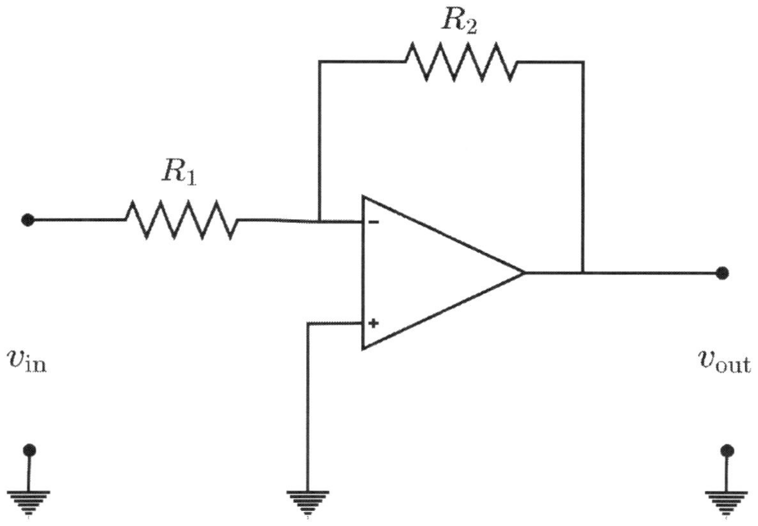


- A. $-\frac{R_1}{R_2}v_{in}$
- B. $-\frac{R_2}{R_1}v_{in}$
- C. $-\frac{R_2}{R_1}$
- D. $\frac{R_2}{R_1}v_{in}$

**58.** A result $R$ is calculated from two measurements as follows $R = 5x_1 - 2x_2^2$. The uncertainty of $x_1$ is $w_1 = \pm 0.01$, and the uncertainty of $x_2$ is $w_2 = \pm 0.02$. What is close to the uncertainty in $R$ for measured values $x_1 = 2$ and $x_2 = 3$ ?

- A. 0.05
- B. 0.16
- C. 0.25
- D. 0.33

**59.** If the switching square wave for a boost converter has a period of 10 µs, the input voltage is 20 V, and the function is ON half of the time, what is the output voltage?

- A. 10 V
- B. 20 V
- C. 30 V
- D. 40 V

**60.** A small business uses an induction motor that requires a complex power of:

$$\mathbf{S}_1 = 30\angle 40^\circ \text{ kV A}$$

and a synchronous motor that requires a complex power of:

$$\mathbf{S}_2 = 15\angle -35^\circ \text{ kV A}$$

The total reactive power (kvar) received by the business is most nearly:

- ○ *A.* 7.91 kvar
- ○ *B.* 15.33 kvar
- ○ *C.* 10.68 kvar
- ○ *D.* 4.27 kvar

**61.** An ammeter, a voltmeter, and a wattmeter are connected to an AC circuit. The devices measure $10\,\text{A}_{\text{rms}}$, $120\,\text{V}_{\text{rms}}$, and $1000\,\text{W}_{\text{rms}}$,respectively. The power factor of this load is most nearly:

- ○ *A.* 0.75
- ○ *B.* 1.2
- ○ *C.* 0.833
- ○ *D.* 0.5

**62.** Three passive loads, $Z_1$, $Z_2$, and $Z_3$, are receiving the complex power values $4+j3$, $5-j$, and $2+j4$V A, respectively. The total complex power received by all loads in parallel with a $50\angle 30^\circ$ voltage source is most nearly:

- ○ *A.* 11-j6 V A
- ○ *B.* -11+j6 V A
- ○ *C.* -11-j6 V A
- ○ *D.* 11+j6 V A

**63.** The line voltage in the wye connected generator that supplies the load in the figure has a peak magnitude of 208 V and are in the positive sequence at 60 Hz. Calculate the phase voltages in instantaneous form $A\cos(\omega t + \phi)$.

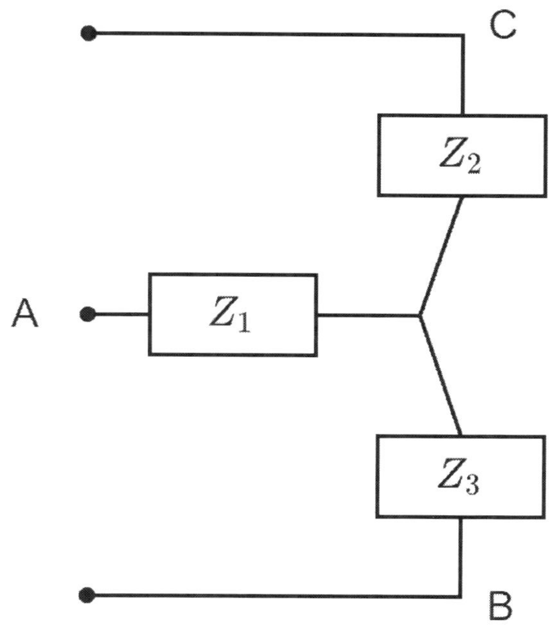


○ A.

$$\mathbf{V}_{an} = 120.09\cos(377t + 30^\circ)\text{ V}$$
$$\mathbf{V}_{bn} = 120.09\cos(377t - 90^\circ)\text{ V}$$
$$\mathbf{V}_{cn} = 120.09\cos(377t + 150^\circ)\text{ V}$$

○ B.

$$\mathbf{V}_{an} = 208\cos(377t + 30^\circ)\text{ V}$$
$$\mathbf{V}_{bn} = 208\cos(377t - 90^\circ)\text{ V}$$
$$\mathbf{V}_{cn} = 208\cos(377t + 150^\circ)\text{ V}$$

○ C.

$$\mathbf{V}_{an} = 120.09\cos(377t + 0^\circ)\text{ V}$$
$$\mathbf{V}_{bn} = 120.09\cos(377t - 120^\circ)\text{ V}$$
$$\mathbf{V}_{cn} = 120.09\cos(377t + 120^\circ)\text{ V}$$

○ D.

$$\mathbf{V}_{an} = 208\cos(377t + 0^\circ)\text{ V}$$
$$\mathbf{V}_{bn} = 208\cos(377t - 120^\circ)\text{ V}$$
$$\mathbf{V}_{cn} = 208\cos(377t + 120^\circ)\text{ V}$$

**64.** A balanced three phase Y connected source has, using a positive phase sequence, $\mathbf{V}_{an} = 120\angle 30° \mathrm{V}_{\mathrm{rms}}$. Find the $\mathbf{V}_{bn}$.

- *A.* $120\angle 90°\ V_{\mathrm{rms}}$
- *B.* $120\angle -90°\ V_{\mathrm{rms}}$
- *C.* $90\angle -90°\ V_{\mathrm{rms}}$
- *D.* $120\angle 30°\ V_{\mathrm{rms}}$

**65.** A load has an effective power of $P = 50\,\mathrm{kW}$ at $400\,\mathrm{V}$, and the power factor is to be compensated from $\cos\theta = 0.8$ to $\cos\theta = 0.95$. Determine the line current after power factor correction.

- *A.* 105.22 A
- *B.* 47.67 A
- *C.* 35.12 A
- *D.* 75.96 A

**66.** A center-tapped transformer has a turns ratio of $4 : 1$ and $120\,\mathrm{V}_{\mathrm{rms}}$ is applied across the primary. What is most nearly the peak voltage across the upper half of the secondary winding?

- *A.* 120 V
- *B.* 10 V
- *C.* 15 V
- *D.* 20 V

**67.** A transformer has $100 : 1$ turns ratio. If 2 A current flows through a $60\,\Omega$ load connected across the secondary, then the applied primary voltage is most nearly:

- *A.* 10 kV
- *B.* 120 kV
- *C.* 12 kV
- *D.* 24 kV

**68.** A permanent magnet DC motor is rated for 25 V, 2 A, and 1300 rpm. If the machine is 90% efficient at rated conditions determine the mechanical power loss if $T_{\text{mechanical loss}} = 0.0334\,\mathrm{N\,m}$.

- *A.* 45.0 W
- *B.* 4.6 W
- *C.* 49.6 W
- *D.* 50.0 W

**69.** Synchronous speed in an AC machine is most nearly defined as:

- *A.* the rotational speed of the motor
- *B.* the rotational speed of the magnetic field
- *C.* the speed at which the torque is maximum
- *D.* the speed at which the machine generates power

**70.** A hollow sphere with four concentrations of charges known to be:

$$Q_1 = +2\,\text{nC}$$
$$Q_2 = +3\text{nC}$$
$$Q_3 = -5\,\text{nC}$$
$$Q_4 = -10\,\text{nC}$$

The electric flux emanating from the sphere is most nearly:

- ○ *A.* 1130 V m
- ○ *B.* 578 V m
- ○ *C.* 0 V m
- ○ *D.* −1130 V m

**71.** Which of the following mathematical expressions of Maxwell's equations is/are false? Select all that apply:

- ☐ *A.* $\nabla \cdot \mathbf{D} = 0$
- ☐ *B.* $\nabla \cdot \mathbf{D} = \rho$
- ☐ *C.* $\nabla \times \mathbf{B} = -\dfrac{\partial \mathbf{B}}{\partial t}$
- ☐ *D.* $\nabla \times \mathbf{E} = -\dfrac{\partial \mathbf{B}}{\partial t}$

**72.** The degree of mismatch in a transmission line is measured by the standing wave ratio, or SWR. What is the range of values that the SWR can have?

- ○ *A.* $-\infty < \text{SWR} \leq 1$
- ○ *B.* $1 \leq \text{SWR} < \infty$
- ○ *C.* $-\infty < \text{SWR} \leq 0$
- ○ *D.* $0 \leq \text{SWR} < \infty$

**73.** A 50 V pulse is incident on a 40 Ω load ($R_L = Z_L$) in a line with characteristic resistance of $Z_0 = R_0 = 60\,\Omega$. Determine the incident current pulse.

- ○ *A.* 0.83 A
- ○ *B.* 1.25 A
- ○ *C.* 0.45 A
- ○ *D.* 1.2 A

**74.** For the system shown in the figure, compute the peak value time:

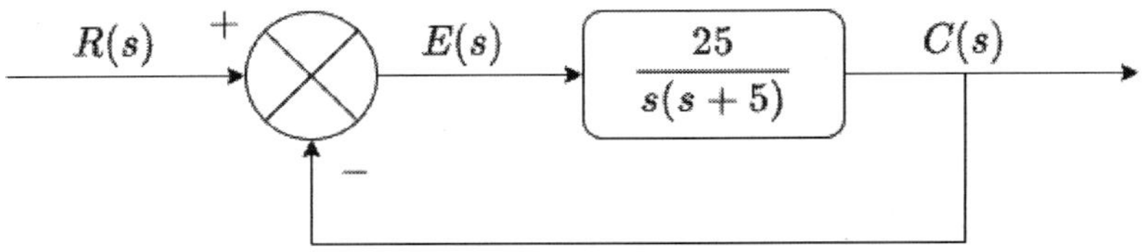


- A. 1.231 s
- B. 0.726 s
- C. 5.111 s
- D. 0.940 s

**75.** A unity feedback control system is shown in the figure. The number of poles in the open-loop characteristic equation for this system is most nearly ________.

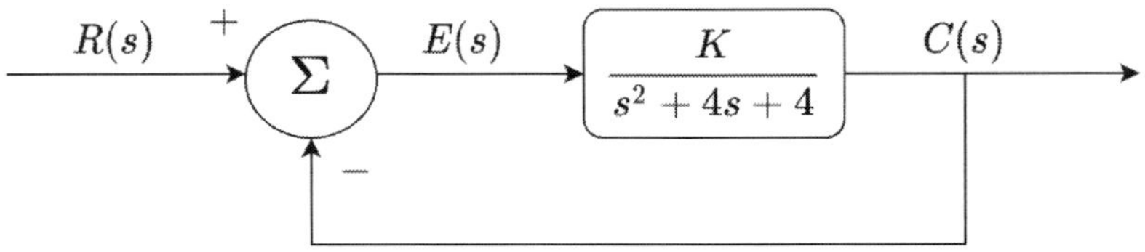


Enter your answer on the blank.

**76.** Consider the plot of the given transfer function, $H(s)$.

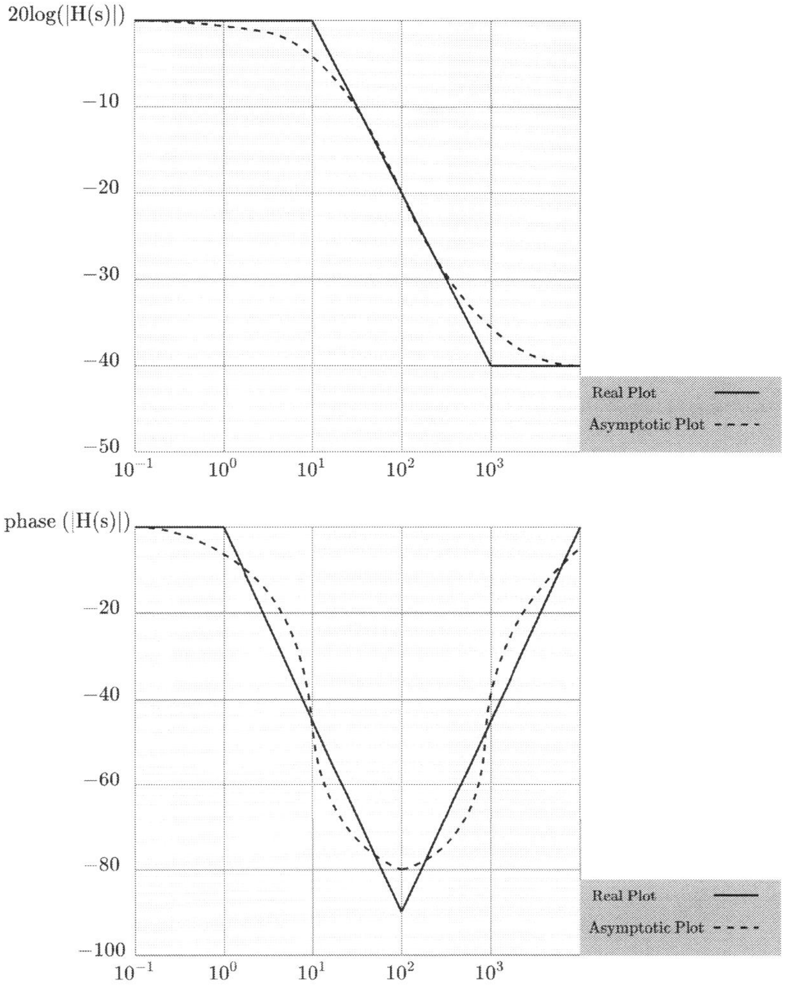


This plot represents a:

- ◯ *A.* band-pass filter
- ◯ *B.* low-pass filter
- ◯ *C.* high-pass filter
- ◯ *D.* band-reject filter

**77.** Consider the transfer function,

$$T(z) = \frac{1}{z^{-2}(1 - 0.9z^{-1})}$$

The finite pole(s) in the $z$-plane is(are) most nearly:

- ○ *A.* z = 1.0; z = 0.9
- ○ *B.* z = 0.9
- ○ *C.* z = 0
- ○ *D.* z = 0.9; z = 0

**78.** Given the transfer function,

$$T(s) = \frac{5s(s+1)}{s^2 + s + 1}$$

Compute the damping ratio.

- ○ *A.* 1/2
- ○ *B.* 0
- ○ *C.* 1
- ○ *D.* 2

**79.** For the system shown in the figure, compute the settling time:

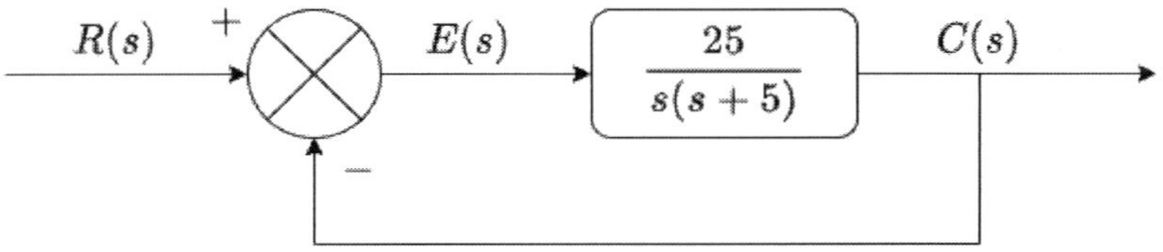


- ○ *A.* 1.6 s
- ○ *B.* 2.4 s
- ○ *C.* 5.2 s
- ○ *D.* 0.2 s

**80.** A TV signal (video and audio) has a bandwidth of 6 MHz. This signal is sampled, quantized, and binary coded to obtain a PCM (pulse code modulated) signal. If the samples are quantized into 1024 levels, how many binary pulses are required to encode each sample?

- ○ *A.* 4
- ○ *B.* 8
- ○ *C.* 10
- ○ *D.* 16

**81.** A digital signal system is required to operate at 9600 bps. If the signal is encoding a 4-bit word, what is the minimum bandwidth of the channel?

- *A.* 1000 Hz
- *B.* 1200 Hz
- *C.* 1400 Hz
- *D.* 1600 Hz

**82.** Determine the Fourier transfom of:

$$f(t) = e^{-t/2}u(t)$$

- *A.* $-\dfrac{1+2j\omega}{2}$
- *B.* $-\dfrac{2}{1+2j\omega}$
- *C.* $\dfrac{1+2j\omega}{2}$
- *D.* $\dfrac{2}{1+2j\omega}$

**83.** Calculate the Fourier transform of the function shown:

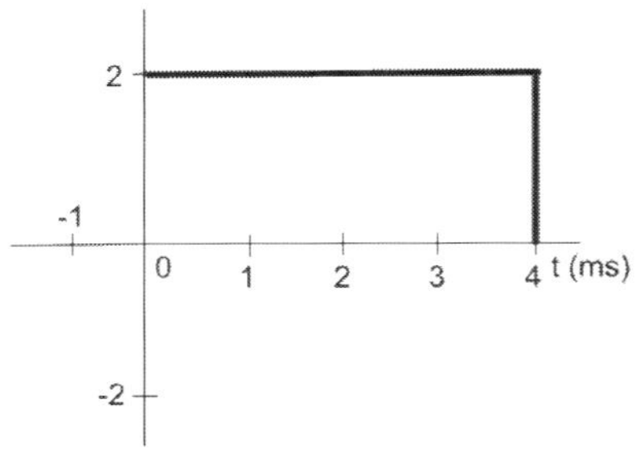


- *A.* $\text{sinc}(4f)e^{-2\pi f}$
- *B.* $8\,\text{sinc}(4f)e^{-4j\pi f}$
- *C.* $8\,\text{sinc}(4f)$
- *D.* $\text{sinc}(f)e^{-4j\pi f}$

**84.** Find the minimum bandwidth required for an FDM communication path using 4 devices, each requiring 3000 Hz and 200 Hz guard band between each device.

- ◯ *A.* 12.6 kHz
- ◯ *B.* 11.4 kHz
- ◯ *C.* 12.0 kHz
- ◯ *D.* 7200 kHz

**85.** When data is transmitted in a serial format, an error detection bit called a parity bit may be appended to the bit stream. Complete the data stream so that the system has odd parity: _ _ `111000`
Select all that apply:

- ☐ *A.* 00
- ☐ *B.* 01
- ☐ *C.* 10
- ☐ *D.* 11

**86.** One of the advantages of digital communications over analog communications is:

- ◯ *A.* digital communications is error free
- ◯ *B.* digital communications do not require encoding
- ◯ *C.* digital communication can withstand channel noise and distortion within limits
- ◯ *D.* digital communications do not require repeaters

**87.** Routing is best defined as:

- ◯ *A.* the process of adding transport headers to data
- ◯ *B.* the process of finding a path on which data can pass from source to destination
- ◯ *C.* the process of filtering errors from data
- ◯ *D.* the process of opening and closing session instances between end-user applications

**88.** Why are terminators required on the main line in bus topology?

- ◯ *A.* To shorten transmission length
- ◯ *B.* To eliminate crosstalk
- ◯ *C.* To eliminate signal bounce
- ◯ *D.* To extend the length of the transmission bus

**89.** DNS stands for:

- ◯ *A.* Designated Network System
- ◯ *B.* Download Network System
- ◯ *C.* Domain Name System
- ◯ *D.* Duplex Number Scheme

**90.** A 33.6 kbps modem across the telephone network fits where in the OSI model?

- ◯ *A.* Data link
- ◯ *B.* Physical
- ◯ *C.* Application
- ◯ *D.* Presentation

**91.** Which of the following are used to evade intrusion detection systems?

- ◯ *A.* Fragmentation
- ◯ *B.* Access attack
- ◯ *C.* Pattern matching
- ◯ *D.* Denial of service

**92.** A network administrator at the PrepFE Headquarters is testing a new security application for monitoring traffic that uses ICMP to multiple host systems. The application is reported as a network attack. This alarm can be referred to as a:

- ◯ *A.* false positive
- ◯ *B.* true positive
- ◯ *C.* false negative
- ◯ *D.* true negative

**93.** A register in a computer contains the following 8-bit value: `00110111`

What is the value of this register when converted into hexadecimal?

- ◯ *A.* $15_{\text{hex}}$
- ◯ *B.* $24_{\text{hex}}$
- ◯ *C.* $37_{\text{hex}}$
- ◯ *D.* $61_{\text{hex}}$

**94.** The decimal equivalent of the binary value `11001` is ________.

Enter your answer on the blank.

**95.** Simplify the expression:

$$(A + C)(AD + A\bar{D}) + AC + C$$

- ◯ *A.* $ABC$
- ◯ *B.* $A + C$
- ◯ *C.* $A + BC$
- ◯ *D.* $A + D$

**96.** The output of a logic gate is '1' when all its inputs are '0'. This gate is either:

- ○ *A.* OR or XNOR
- ○ *B.* NAND or XOR
- ○ *C.* NOR or XNOR
- ○ *D.* AND or XOR

**97.** Find the minimal sum of products expression for the function shown in the mapping:

| C \ AB | 00 | 01 | 11 | 10 |
|---|---|---|---|---|
| 0 | 0 | 0 | 0 | 0 |
| 1 | 0 | 1 | 1 | 1 |

- ○ *A.* $F = AB + AC$
- ○ *B.* $F = BC$
- ○ *C.* $F = AC + BC$
- ○ *D.* $F = AB + BC$

**98.** What states do $Q$ and $\overline{Q}$ toggle to for the D flip-flop during the second B pulse?

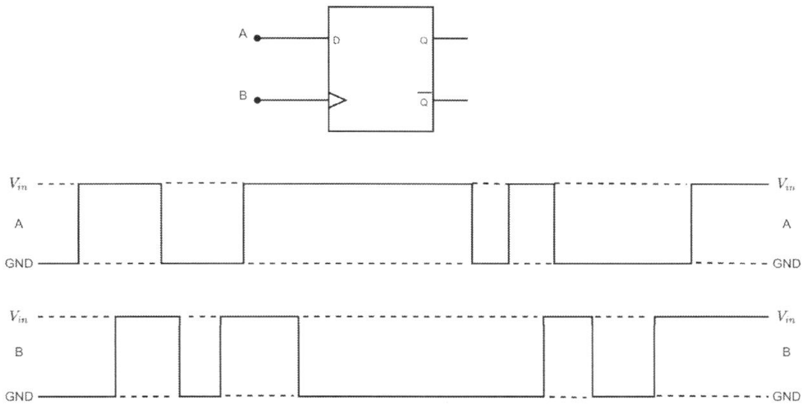


- A. $Q = 1$
  $\overline{Q} = 1$
- B. $Q = 1$
  $\overline{Q} = 0$
- C. $Q = 0$
  $\overline{Q} = 0$
- D. $Q = 0$
  $\overline{Q} = 1$

**99.** Given the programming circuit below for a PLA, what is the logical implementation at Output Y?

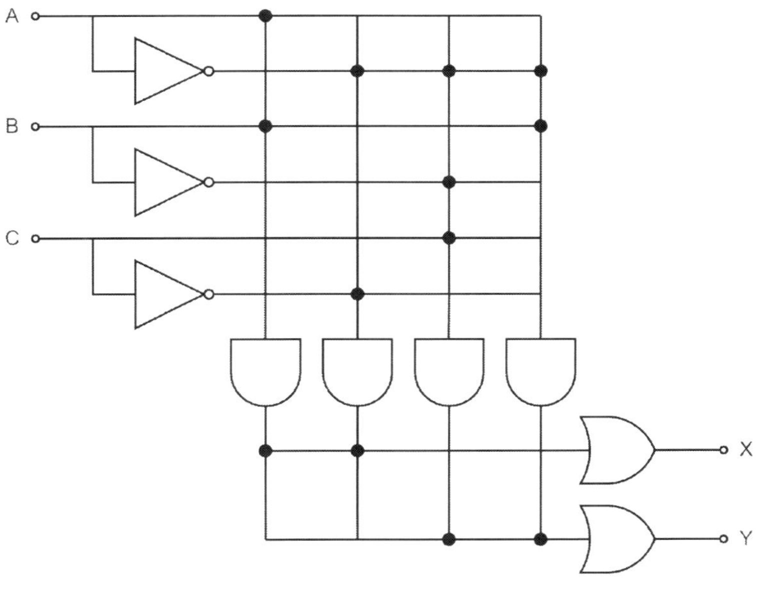


- A. $AC + \bar{A}B\bar{C}$
- B. $\bar{A}BC + ABC + \bar{A}\bar{C}$
- C. $AB + \bar{A}\bar{C}$
- D. $\bar{A}\bar{B}C + \bar{A}B$

**100.** The state diagram of a Mealy machine is shown below.
The transitions between states are denoted with direct lines. The expressions x/y denote input/output. What is the input/output combination that corresponds to a state transition from B to C?

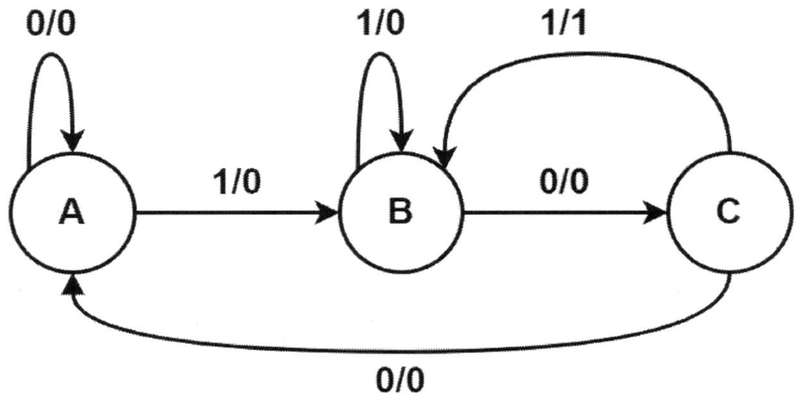


- A. 1/0
- B. 0/0
- C. 0/1
- D. 1/1

**101.** Consider the implementation of the logic function, $A \cdot \bar{A} = Z$, shown below. Theoretically the output will always be 0.
If the output from the inverter is delayed, when the input $A$ transitions from 0 to 1, what will happen to the output $Z$?

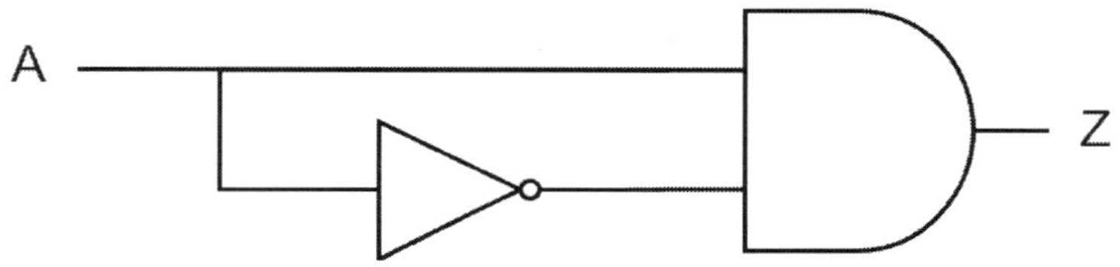


- A. Output $Z$ will toggle to logic 0 and stay there
- B. Output $Z$ will glitch to 0 for a moment
- C. Output $Z$ will toggle to logic 1 and stay there
- D. Output Z will glitch to logic 1 for a moment

**102.** The bus that chooses a location for storing information within a computer system is the:

- A. control bus
- B. data bus
- C. address bus
- D. main bus

**103.** Which of these is NOT a flag in an 8086 microprocessor?

- ◯ *A.* Parity
- ◯ *B.* Carry
- ◯ *C.* Division
- ◯ *D.* Sign

**104.** Which of the following is true about EPROM?

- ◯ *A.* EPROM can be written to only once
- ◯ *B.* EPROM is programmed with fuses
- ◯ *C.* EPROM can be electrically erased
- ◯ *D.* EPROM can be erased by UV light

**105.** Which of the following is the correct order for memory speed from fastest to slowest?

- ◯ *A.* Hard disk, cache memory, CD-ROM, main memory
- ◯ *B.* Cache memory, main memory, hard disk, CD-ROM
- ◯ *C.* Main memory, cache memory, hard disk, CD-ROM
- ◯ *D.* CD-ROM, hard disk, main memory, cache memory

**106.** Which of these are types of Interfacing in a microprocessor?
Select all that apply:

- ☐ *A.* The path of communication between two components
- ☐ *B.* Accessing memory
- ☐ *C.* The method a user employs to interact with the components
- ☐ *D.* None of these

**107.** An array consists of n elements. We want to create a heap using the elements. The time complexity of building a heap will be in order of:

- ◯ *A.* $\mathrm{O}(n \cdot \log n)$
- ◯ *B.* $\mathrm{O}(n \cdot n \cdot \log n)$
- ◯ *C.* $\mathrm{O}(n \cdot n)$
- ◯ *D.* $\mathrm{O}(n \cdot \log n \cdot \log n)$

**108.** Which of the following is NOT a standard operation on a skip list?

- ◯ *A.* Search Operation
- ◯ *B.* Coupling Operation
- ◯ *C.* Deletion Operation
- ◯ *D.* Insertion Operation

**109.** What is the main disadvantage of implementing a tree using fixed arrays?

- ○ *A.* Must know the maximum number of nodes possible before creating the trees
- ○ *B.* Difficult to find the parent of a node
- ○ *C.* Difficult to implement
- ○ *D.* Difficulty in knowing children nodes of a node

**110.** Given the following pseudocode, what is nearest the final value of $i$?

```
Set i = 13
Set j = 4
While j <= 6
   i = i/j
   j = j+2
End While
Print
```

- ○ *A.* 0.25
- ○ *B.* 0.54
- ○ *C.* 0.77
- ○ *D.* 1.24

# Solutions

| | | | |
|---|---|---|---|
| 1. C | 29. D | 57. B | 85. A, D |
| 2. D | 30. C | 58. C | 86. C |
| 3. B | 31. D | 59. D | 87. B |
| 4. C | 32. B | 60. C | 88. C |
| 5. 689 | 33. 107 Ω | 61. C | 89. C |
| 6. A | 34. A | 62. D | 90. A |
| 7. D | 35. D | 63. C | 91. A |
| 8. A | 36. A | 64. B | 92. A |
| 9. C | 37. D | 65. D | 93. C |
| 10. 2 | 38. D | 66. D | 94. 25 |
| 11. D | 39. A | 67. C | 95. B |
| 12. 0.55, 0.61, 0.66 | 40. C | 68. B | 96. C |
| 13. D | 41. A | 69. B | 97. C |
| 14. C | 42. C | 70. D | 98. B |
| 15. A | 43. B | 71. C | 99. D |
| 16. D | 44. A | 72. B | 100. B |
| 17. D | 45. B | 73. A | 101. D |
| 18. A | 46. C | 74. B | 102. C |
| 19. D | 47. A | 75. 2 | 103. C |
| 20. D | 48. B | 76. B | 104. D |
| 21. C | 49. A | 77. B | 105. B |
| 22. A | 50. B | 78. A | 106. A, B, C |
| 23. A | 51. A | 79. A | 107. A |
| 24. B | 52. D | 80. C | 108. B |
| 25. D | 53. C | 81. B | 109. A |
| 26. D | 54. D | 82. D | 110. B |
| 27. A | 55. B | 83. B | |
| 28. C | 56. A | 84. A | |

You can find detailed solutions beginning on the next page.

**1.** **THE CORRECT ANSWER IS: C**

2, -5

---

Refer to the Quadratic Equation section in the Mathematics chapter of the *FE Reference Handbook.*

We know the polynomial in the problem statement is quadratic, as opposed to linear or cubic, as the formula in the problem statement can be rearranged to standard quadratic form such as

$$(a)x^2 + (b)x + (c) = 0$$
$$x^2 + 3x - 10 = 0$$

Now that our equation is in standard quadratic form, we can use the quadratic equations to solve for $x$.

$$x = \frac{-b \pm \sqrt{b^2 - 4ac}}{2a}$$

and

$$(a)x^2 + (b)x + (c) = 0$$

The quadratic formula in our problem then becomes

$$x^2 + 3x - 10 = 0$$

and

$$x = \frac{-(3) \pm \sqrt{(3)^2 - 4(1)(-10)}}{2(1)}$$
$$x = 2 \text{ and } x = -5$$

**2.** **THE CORRECT ANSWER IS: D**

$10.82\angle 56.31°$

---

Refer to the Algebra of Complex Numbers & Polar Coordinate System section in the Mathematics chapter of the *FE Reference Handbook.*

The radius and angle of the polar coordinates can be determined from the given x and y coordinates. These values are given by,

$$r = \sqrt{x^2 + y^2}$$
$$\theta = \arctan \frac{y}{x}$$

Substituting the known values to find the radius,

$$\begin{aligned} r &= \sqrt{6^2 + 9^2} \\ &= \sqrt{36 + 81} \\ &= \sqrt{117} \\ &\approx 10.82 \end{aligned}$$

Then solving for the angle, $\theta$,

$$\theta = \arctan\frac{9}{6} = 56.31^\circ$$

Therefore, rectangular coordinates $(6, 9) \approx 10.82\angle 56.31^\circ$ in polar coordinates

**3.** **THE CORRECT ANSWER IS: B**

$$\{(1,2)(1,4)(3,2)(3,4)\}$$

---

Refer to the Discrete Math section in the Mathematics chapter of the *FE Reference Handbook.*

The Cartesian product is defined as the set of all ordered pairs $(a, b)$ where all elements of $a$ are in A, and all elements of $b$ are in B. In set notation, this is written as,

$$A \times B = \{(a,b)|a \in A \text{ and } b \in B\}$$

Apply this definition to the given sets to obtain,

$$A \times B = \{(1,2)(1,4)(3,2)(3,4)\}$$

**4.** **THE CORRECT ANSWER IS: C**

$$\{1,5\}$$

---

Refer to Discrete Math section in the Mathematics chapter of the *FE Reference Handbook.*

The symbol $\cap$ means "intersection" which denotes all the elements of one set that also belong to another set. In other words, all the elements of $A$ that belong to $B$, and vice versa.

The only elements that are in $A$ and $B$ are 1 and 5.

The intersection of $A$ and $B$, $A \cap B$, is therefore,

$$A \cap B = \{1,5\}$$

**5.** **THE CORRECT ANSWER IS: 689**

---

Refer to the Straight Line section in the Mathematics chapter of the *FE Reference Handbook.*

The distance between two points is

$$d = \sqrt{(y_2 - y_1)^2 + (x_2 - x_1)^2}$$
$$d = \sqrt{(168 - 315)^2 + (740 - 67)^2}$$
$$d = 688.87 \approx 689$$

**6.** **THE CORRECT ANSWER IS: A**

-1.27

---

Refer to the Indefinite Integrals section in the Mathematics chapter of the *FE Reference Handbook.*

The integral of a function $f(x)$ gives you the area under the curve formed by that function.

Integrate the curve from $x = 0$ to $x = 1.5$

$$\begin{aligned} A &= \int_0^{1.5} -x^3 \, \mathrm{d}x \\ &= -\frac{x^4}{4}\bigg|_0^{1.5} \\ &= -1.27 \end{aligned}$$

Real world area values are scalar i.e. positive. A negative 1.27 area value, such as the one in this problem, indicates the area occurs under the x axis.

**7.** **THE CORRECT ANSWER IS: D**

$$5\sec^2(5x)$$

---

Refer to the Derivatives section in the Mathematics chapter of the *FE Reference Handbook*. There, you will find a long list of typical derivatives.

According to this list, the derivative of $\tan(u)$ is $\sec^2(u)$, and the derivative of $cu$ is $c$, where $u$ is a variable and $c$ is a number.

If we take the two derivative identities stated above and apply them to $\tan(5x)$, we get:

$$\frac{\mathrm{d}}{\mathrm{d}x}(\tan(5x)) = 5\sec^2(5x)$$

**8.** **THE CORRECT ANSWER IS: A**

$$x(t) = Ce^{5t} + \tfrac{3}{5}$$

---

Refer to the Differential Equations section in the Mathematics chapter of the *FE Reference Handbook*.

Begin by multiplying both sides of the equation by the differential operator $\mathrm{d}t$ and dividing by $5x - 3$ to collect all variable terms to each side of the equation.

$$\frac{\mathrm{d}x}{5x-3} = \mathrm{d}t$$

Now, integrate both sides. Refer to the table of Indefinite Integrals section in the Mathematics chapter of the *FE Reference Handbook*.

$$\begin{aligned} \int \frac{\mathrm{d}x}{5x-3} &= \int \mathrm{d}t \\ \frac{1}{5}\log|5x-3| &= t + C \\ \log|5x-3| &= 5t + 5C \end{aligned}$$

Now, exponentiate both sides to eliminate the logarithm,

$$5x - 3 = \pm e^{(5t+5C)}$$
$$x = \pm\frac{1}{5}e^{(5t+5C)} + \frac{3}{5}$$
$$x = \pm\frac{1}{5}e^{5C}e^{5t} + \frac{3}{5}$$

Where we make use of the identity $e^{x+y} = e^x e^y$ to help simplify. Further, we note that $\pm\frac{1}{5}e^{5C}$ is still a constant which we just rename $C$, so the new expression for $x(t)$ can be written as,

$$x(t) = Ce^{5t} + \frac{3}{5}$$

The value of $C$ would be determined by initial conditions if they were given by the problem statement.

**9.** **THE CORRECT ANSWER IS: C**

$$\frac{1}{x} + \frac{1}{y} = C$$

---

Refer to the Differential Equations section in the Mathematics chapter of the *FE Reference Handbook*.

First, write the term $y'$ in differential operator notation,

$$y^2 + x^2\frac{\mathrm{d}y}{\mathrm{d}x} = 0$$

Now collect terms,

$$\frac{\mathrm{d}y}{\mathrm{d}x} = -y^2\frac{1}{x^2}$$

Next, multiply both sides by their respective differential operators to obtain an expression that can be integrated,

$$\frac{1}{y^2}\,\mathrm{d}y = -\frac{1}{x^2}\,\mathrm{d}x$$
$$\int \frac{1}{y^2}\,\mathrm{d}y = -\int \frac{1}{x^2}\,\mathrm{d}x$$

The integrated expression can be written as,

$$-\frac{1}{y} = \frac{1}{x} - C$$

Note that the constant of integration has been selected to be $-C$ out of convenience so that the final general solution can be written directly as,

$$\frac{1}{x} + \frac{1}{y} = C$$

Ultimately the sign selection for $C$ does not matter since it is only a constant of integration - arithmetic manipulation will arrive at this general form.

**10. THE CORRECT ANSWER IS: 2**

---

Refer to the Matrices section in the Mathematics chapter of the *FE Reference Handbook.*

The rank of a matrix is equal to the number of rows that are linearly independent. There are many ways to manipulate a matrix to determine the linear dependence of rows. We will show one such methodical approach. Your approach may be different but should arrive at the same conclusion.

Multiply row-2 by 2, and subtract it from row-3, giving:

$$\begin{bmatrix} 3 & -4 & -6 \\ 2 & 4 & -4 \\ 0 & -2 & 0 \end{bmatrix}$$

Next we multiply row-2 by 3, and row-1 by 2, then subtract row-1 from row-2, which gives:

$$\begin{bmatrix} 3 & -4 & -6 \\ 0 & 20 & 0 \\ 0 & -2 & 0 \end{bmatrix}$$

At this point we can already see that row-2 and row-3 are linearly dependent (they only vary by a multiplicative factor) but we will continue with the methodical removal. Finally we just divide row-1 by 3, row-2 by 20, and row-3 by -2 to put in a simplified form:

$$\begin{bmatrix} 1 & -4/3 & -2 \\ 0 & 1 & 0 \\ 0 & 1 & 0 \end{bmatrix}$$

Thus the last step is to subtract row-2 from row-3, resulting in:

$$\begin{bmatrix} 1 & -4/3 & -2 \\ 0 & 1 & 0 \\ 0 & 0 & 0 \end{bmatrix}$$

and from this form there is nothing else we can do to reduce the matrix, and we can see there are 2 linearly independent rows (2 non-zero rows), and thus the rank is 2.

**11. THE CORRECT ANSWER IS: D**

$$x^3 + 6xy^2z + 5y$$

---

Refer to the Gradient, Divergence, and Curl section in the Mathematics chapter of the *FE Reference Handbook.*

The divergence of a vector field is defined as:

$$\begin{aligned} \nabla \cdot \mathbf{V} &= \left( \frac{\partial}{\partial x}\mathbf{i} + \frac{\partial}{\partial y}\mathbf{j} + \frac{\partial}{\partial z}\mathbf{k} \right) \cdot (V_1\mathbf{i} + V_2\mathbf{j} + V_3\mathbf{k}) \\ &= \frac{\partial}{\partial x}V_1 + \frac{\partial}{\partial y}V_2 + \frac{\partial}{\partial z}V_3 \end{aligned}$$

The dot product reduces the expression to a scalar, and we are left with the derivatives with respect to each variable. Therefore, the divergence of the vector field **D** is

$$\begin{aligned}\nabla \cdot \mathbf{D} &= \frac{\partial}{\partial x}(5xy) + \frac{\partial}{\partial y}(2xy^3z) + \frac{\partial}{\partial z}(x^3z) \\ &= 5y + 6xy^2z + x^3 \\ &= x^3 + 6xy^2z + 5y\end{aligned}$$

**12. THE CORRECT ANSWER IS: Mode of 0.55; median of 0.61; mean of 0.66**

---

Refer to the Dispersion, Mean, Median, and Mode Values in the Engineering Probability and Statistics chapter of the *FE Reference Handbook.*

The *mode* of a dataset is the value that occurs with the greatest frequency, for the given dataset this is seen to be *mode*= 0.55.

Before you can calculate the *median* of a dataset, you must rearrange the dataset in increasing order.

$$0.55, 0.55, 0.60, 0.61, 0.74, 0.76, 0.80$$

When the discrete data are rearranged in increasing order, and n is odd, the median is the value of the $(\frac{n+1}{2})^{th}$ item.

$$\left[\frac{n+1}{2}\right]^{th} = \left[\frac{7+1}{2}\right]^{th} = 4^{\text{th}}$$

The $4^{\text{th}}$ value in the ordered dataset is 0.61

The above is the *FE Reference Handbook* way of determining the median. The easy way to find the median is to find the number that is in the middle of an **ordered** list of numbers. In instances where you have two numbers in the middle of the ordered list, the median will be the average of the two numbers in the middle.

The *mean* of the dataset (generally speaking, mean is implied to mean *arithmetic mean* and not *geometric mean* or *weighted arithmetic mean*) is found from

$$\begin{aligned}\bar{X} &= \frac{1}{n}\sum_{i=1}^{n} X_i \\ &= \frac{1}{7}(0.61 + 0.74 + 0.76 + 0.55 + 0.80 + 0.55 + 0.60) \\ &= 0.66\end{aligned}$$

**13. THE CORRECT ANSWER IS: D**

88%

---

Ultimately, we must determine the probability of flipping heads at least one time, $P_{1h}$, after flipping a coin three times but is found most easily from reconsidering the problem as the probability of *not flipping a tail all three times*, or $P_{1h} = 1 - P_{3t}$

We know in a given coin flip, we have a 0.5 (50%) probability of flipping tails. Therefore, flipping tails every toss for three tosses is

$$P_{3t} = 0.5 \times 0.5 \times 0.5 = 0.125 \Rightarrow 12.5\%$$

Thus

$$P_{1h} = 1 - 0.125 = 0.875 \Rightarrow 87.5\%$$

**14.** **THE CORRECT ANSWER IS: C**

84%

---

Refer to the Law of Compound or Joint Probability section in the Engineering Probability and Statistics chapter of the *FE Reference Handbook.*

Ultimately, we must determine the probability of someone using the app for 1 month given s/he uses the app for 1 week.

Let's assign metric #1 as A and metric #2 as B.

$$P(A) = \text{probability of using app for 1 week} = 0.95$$
$$P(B) = \text{probability of using app for 1 month} = 0.80$$

Therefore:

$P(A|B)$ is probability of A happening given that B occurs i.e. the probability someone uses the app for 1 week given s/he used the app for 1 month.

$P(B|A)$ is probability of B happening given that A occurs i.e. the probability someone uses the app for 1 month given s/he used the app for 1 week.

Ultimately, we must determine the probability of B happening given A occurs i.e. $P(B|A)$. We can use the following equation to calculate $P(B|A)$.

$$P(A)P(B\,|\,A) = P(B)P(A\,|\,B)$$

then solving for $P(B|A)$

$P(A|B)$ is 1.0 (or 100%) probability because any user who used the app for 1 month, by default, used the app for 1 week.

Now that we know all the unknowns in the equation, we can solve for $P(B|A)$.

$$P(A)P(B\,|\,A) = P(B)P(A\,|\,B)$$
$$0.95\,[P(B|A)] = 0.80 \cdot 1.0$$

solve for $P(B|A)$,

$$P(B|A) = 0.84 \text{ or } 84\%$$

So, the probability someone becomes a 1-month user after using the app for 1 week is 84%.

**15.** **THE CORRECT ANSWER IS: A**

0.34

---

Refer to the Probability Functions, Distributions, and Expected Values section in the Engineering Probability and Statistics chapter of the *FE Reference Handbook.*

We identify this problem's probability as a discrete distribution as the number of possible solutions is finite. The number of polled students who approve of 8AM classes in our problem could be 0 students, 1 student, or 2 students. Additionally, we identify this problem as a discrete Bernoulli distribution as there are only two possible outcomes: True or False. True is when the student approves of 8AM classes. False is when the student does not approve of 8AM classes.

List out the possible results every time we randomly select two polled students, and we ask them if they approve or do not approve of 8AM classes:

Scenario A $\rightarrow$ Student 1: Approve; Student 2: Approve

Scenario B $\rightarrow$ Student 1: Approve; Student 2: Do not approve

Scenario C $\rightarrow$ Student 1: Do not approve; Student 2: Approve

Scenario D $\rightarrow$ Student 1: Do not approve; Student 2: Do not approve

Calculate the probability of each scenario from occurring.

Scenario A $\rightarrow$ Student 1: Approve; Student 2: Approve

The probability of 1 student approving 8AM classes is 22% (or 0.22). Therefore, the probability of the first student approving, and the second student approving 8AM classes is:

$$0.22 \times 0.22 = 0.0484 = 4.84\%$$

Scenario B $\rightarrow$ Student 1: Approve; Student 2: Do not approve

If the probability of 1 student approving 8AM classes is 22%, then probability of a student not approving 8AM classes is the opposite. The probability of 1 student not approving 8AM classes is $100\% - 22\% = 78\% = 0.78$. Therefore, the probability of the first student approving, and the second student not approving is:

$$0.22 \times 0.78 = 0.1716 = 17.16\%$$

Scenario C $\rightarrow$ Student 1: Do not approve; Student 2: Approve

If the probability of 1 student approving 8AM classes is 22%, then probability of a student not approving 8AM classes is the opposite. The probability of 1 student not approving 8AM classes is $100\% - 22\% = 78\% = 0.78$. Therefore, the probability of the first student not approving, and the second student approving 8AM classes is:

$$0.78 \times 0.22 = 0.1716 = 17.16\%$$

Scenario D $\rightarrow$ Student 1: Do not approve; Student 2: Do not approve

The probability of the first student not approving, and the second student not approving is:

$$0.78 \times 0.78 = 0.6084 = 60.84\%$$

| # of students out of 2 | 0 students approving | 1 student approving | 2 students approving |
|---|---|---|---|
| P(x) | 60.84% | 2*17.16% =34.32% | 4.84% |

Now, onto the actual problem solution:

Expected value mean is simply calculated by multiplying each of the possible outcomes by the likelihood each outcome will occur, and then summing all of those values.

$$\begin{aligned}\mu = E[X] &= \sum_{k=1}^{n} x_k f(x_k) \\ &= 0(0.6084) + 1(0.3432) + 2(0.0484) \\ &= 0.44\end{aligned}$$

The expected value variance can be calculated as

$$\begin{aligned}\sigma^2 = V[X] &= \sum_{k=1}^{n} (x_k - \mu)^2 f(x_k) \\ &= (0 - 0.44)^2 \cdot 0.6084 + (1 - 0.44)^2 \cdot 0.3432 + (2 - 0.44)^2 \cdot 0.0484 \\ &= 0.3432\end{aligned}$$

**16.** **THE CORRECT ANSWER IS: D**

liability insurance

It is not possible to directly search this in the *FE Reference Handbook*. But the FE exam will expect a certain level of knowledge of the engineer that exists outside of the handbook, and a problem such as this is fair game. Liability insurance is coverage for bodily injury and property damage claims made by others. It includes general liability insurance, auto liability insurance, workers compensation insurance, *etc*.

**17.** **THE CORRECT ANSWER IS: D**

You apply for PE licensure in Florida

Refer to the Board and Jurisdiction - Model Law, Section 110.20 Definitions section in the Ethics chapter of the *FE Reference Handbook*.

A professional engineer (PE) is only allowed to sign and stamp engineering documents for the state in which s/he is licensed. You may be licensed in more than one state if desired. As a PE, you have to be licensed in the state where your design is going to be **built**.

Additional facts about licensure: For some states, you just have to file paperwork to get licensed in that state in addition to your home state - no need to take more PE tests. However, states like California might have additional technical-regulatory requirements not applicable to your home state (e.g. seismic considerations - Geotechnical PE and Structural PE). Due to these additional technical and regulatory requirements, additional examination will be required if one wishes to be licensed in that new state in addition to the home state.

**18.** **THE CORRECT ANSWER IS: A**

Yes, if they file the patent within 12 months of the public demonstration

---

According to US Patent Law, 35 U.S.C. 102, a patent may still be granted even with public disclosure if the effective filing date for the patent is within 12 months of the disclosure.

It is important to note that patents are on a 'first-to-file' basis so the investors to whom the invention was demonstrated could file their own patent on the invention if they do so within 12 months of the disclosure and before the original inventor files theirs.

**19.** **THE CORRECT ANSWER IS: D**

All of these

---

Refer to the Safety and Prevention section in the Safety chapter of the *FE Reference Handbook.*

The answer is not directly searchable but engineers should be familiar with PPE and scenarios that may require it. Safety rated glasses are important to wear as the following hazards may be encountered around electrical circuits:

- Arc flash
- Soldering fumes
- Sparks that may cause an explosion in the vicinity of explosive materials

All of these and more are reasons to wear safety glasses while working on electrical circuits.

**20.** **THE CORRECT ANSWER IS: D**

$6,360

---

Refer to the Engineering Economics chapter of the *FE Reference Handbook.* MACRS depreciation can be found from the following equation:

$$D_j = C(\text{factor})$$

Refer to the MACRS table in the Engineering Economics chapter of the *FE Reference Handbook* to obtain the "factor" in the $D_j$ formula.

Using a MACRS table, the depreciation value at **1 year of depreciation** for a piece of equipment that has a service life of 10 years is:

$$D_{year1} = \left[(\$15,000)(\frac{10.00\%}{100})\right] = \$1,500$$

Using a MACRS table, the depreciation value at **2 years of depreciation** for a piece of equipment that has a service life of 10 years is:

$$D_{year2} = \left[(\$15,000)(\frac{18.00\%}{100})\right] = \$2,700$$

Using a MACRS table, the depreciation value at **3 years of depreciation** for a piece of equipment that has a service life of 10 years is:

$$D_{year3} = \left[(\$15,000)(\frac{14.40\%}{100})\right] = \$2,160$$

Total depreciation after 3 years:

$$\sum D_j = \$1,500 + \$2,700 + \$2,160 = \$6,360$$

**21.** **THE CORRECT ANSWER IS: C**

$13,236

---

Refer to the formula table in the Engineering Economics chapter of the *FE Reference Handbook*.

Ultimately, we must calculate the future worth (F) given a present value (P) and then subtract the initial account deposit to determine how much money in the account balance came from the CD interest rate alone.

$$(F/P, i\%, n) \rightarrow F = P(1+i)^n$$

We must "annualize" the interest rate since the problem statement indicates interest compounding happens semiannually and not annually. Annualize the interest rate:

$$i_e = \left(1 + \frac{r}{m}\right)^m - 1 = \left(1 + \frac{0.021}{2}\right)^2 - 1 = 0.0211$$

$m = 2$ because there are 2 compounding periods in one year of semiannually compounding. Had compounding been e.g. monthly, $m$ would have been equal to 12.

We calculate the future worth (F) now that we have an annualized interest rate.

$$F = P(1+i)^n = \$57,000(1+0.0211)^{10} = \$70,235.6$$

Subtract the future worth from the initial deposit to determine how much was earned from just the CD account interest after 10 years.

$$\$70,235.6 - \$57,000 = \$13,236$$

**22.** **THE CORRECT ANSWER IS: A**

Remodeling the home and renting it for $2,800/month

---

Refer to the Rate-of-Return and Benefit-Cost Analysis sections in Engineering Economics chapter of the *FE Reference Handbook*.

We must calculate and then compare the rate-of-return interest for the non-remodeled home and the remodeled home.

The rate-of-return interest refers to the interest rate, $i$, that will return initial investments, or in other words, when the benefits and costs are equal, $B - C = 0$.

The costs, $C$, are the purchase price and remodeling cost. The benefit, $B$, is the uniform rent at some interest rate over 15 years. The trick here is that we must know that the benefit is the *Present Worth* of the 15 years of rent to compare it with the upfront costs.

Referring to the factor table from the *FE Reference Handbook*, we must find present worth, $P$, given a uniform amount per interest period, $A$ (*i.e.* to $P$ given $A$). The symbol for this factor is $(P/A, i\%, n)$. The formula for this factor is,

$$(P/A, i\%, n) = \frac{(1+i)^n - 1}{i(1+i)^n}$$

**Calculate the rate-of-return for the non-remodeled home.** There are two practical approaches here:

1. Use the Interest Rate Tables from the *FE Reference Handbook* to look up factor values.
2. Use the equation solver on your calculator to solve for $i$ directly.

To convert from $A$ to $P$, we have:

$$P = A(P/A, i\%, n)_{rent} = A\frac{(1+i)^{15} - 1}{i(1+i)^{15}}$$

where $A = \$2,000$ month $\times$ 12 months $= \$24,000$ annual rent.

Substituting this into our benefit-cost equation, $B - C = 0$:

$$\begin{aligned} A(P/A, i\%, n)_{rent} - C &= 0 \\ \$24,000(P/A, i\%, n)_{rent} - \$250,000 &= 0 \\ (P/A, i\%, n)_{rent} &\approx 10.4 \end{aligned}$$

We have recreated a portion of several of the *Interest Rate Tables* below to simulate the look-up.

Factor Table - i = 4.00%

| n | P/F | P/A | P/G |
|---|---|---|---|
| 14 | 0.5775 | 10.5631 | 61.9618 |
| 15 | 0.5553 | 11.1184 | 69.7355 |
| 16 | 0.5339 | 11.6523 | 77.7441 |

Factor Table - i = 6.00%

| n | P/F | P/A | P/G |
|---|---|---|---|
| 14 | 0.4423 | 9.2950 | 51.7128 |
| 15 | 0.4173 | 9.7122 | 57.5546 |
| 16 | 0.3936 | 10.1059 | 63.4592 |

Factor Table - i = 8.00%

| n | P/F | P/A | P/G |
|---|---|---|---|
| 14 | 0.3405 | 8.2442 | 43.4723 |
| 15 | 0.3152 | 8.5595 | 47.8857 |
| 16 | 0.2919 | 8.8514 | 52.2640 |

In the first table above ($i = 4.00\%$), we go down column $n$ to look for 15, which was given in the problem statement as the expected time for return. Then we move along the row of $n = 15$ to the column for $P/A$, to find the value 11.1184. Note that this value is larger than the value of 10.4 by about 0.7.

So, next we go to the table for $i = 6.00\%$. Again looking for $n = 15$ and $P/A$, the value is 9.7122. This value is lower than 10.4 by about 0.7.

At this point, to be accurate we would want to run a full interpolation of these lookup values in order to find a more-precise rate-of-return. Even in the age of calculators, table interpolation is still a very powerful tool for engineers! However, we are taking a timed-exam, so we are free to take some shortcuts and will skip a formal interpolation for now and just estimate the rate-of-return, $i$.

Because we saw that the value of $P/A$ for $i = 4\%$ is *larger* than our desired value by 0.7, and that similarly $i = 6\%$ is *smaller* by 0.7, it is perfectly reasonable to estimate that the rate-of-return for the non-remodeled home, with value of $P/A = 10.4$, is $i \approx 5\%$.

Alternatively, if instead of looking at any of these tables, we wanted to plug in the formula for $(P/A, i\%, n)$ we have,

$$(P/A, i\%, n)_{rent} \approx 10.4$$
$$\frac{(1+i)^n - 1}{i(1+i)^n} = 10.4$$

This yields an algebraic equation which is not reasonable to solve by hand. If you are comfortable with your NCEES®-approved calculator which can handle algebraic solutions, you could plug this equation in and find a value of $i$,

$$i = 0.0494 \rightarrow 4.9\%$$

Which we note is very close to the $i = 5\%$ we approximated with a couple, very quick table look-ups!

**Calculate the rate-of-return for the remodeled home.** We can follow all the same steps as above for the non-remodeled home but with slightly different investment and rent amounts,

$$A(P/A, i\%, n)_{rent} - C = 0$$

where $A = \$2,800$ month $\times$ 12 months $= \$33,600$ annual rent. Our benefit-cost equation for the remodeled home is then,

$$\$33,600(P/A, i\%, n)_{rent} - (\$250,000 + \$70,000) = 0$$

Thus,

$$(P/A, i\%, n)_{rent} \approx 9.5$$

Because we already looked up values in the table for $i = 6\%$, we know $P/A = 9.7122$. This is slightly larger than our value of 9.5. Looking at the table for $i = 8\%$, we find $P/A = 8.5595$. Again, instead of running a full interpolation it suffices to observe that our value $P/A = 9.5$ is somewhere between 6% and 8% (though much closer to $i = 6\%$).

For this problem, what we are ultimately asked to determine is which of the two choices (don't remodel or remodel) result in a larger rate-of-return. Because we don't need to know *exactly how much larger* one is from the other, it isn't worth our time to determine precisely what the effective interest rate is. Instead it is enough to recognize that for the remodeled home, $i > 6.00\%$ which is already larger than the rate-of-return for the non-remodeled home.

However, if we did have access to a calculator that can handle algebraic solutions, we could bypass the table look-ups and solve for $i$ just as we had for the non-remodeled home,

$$\begin{aligned}(P/A, i\%, n)_{rent} &\approx 9.5 \\ \frac{(1+i)^n - 1}{i(1+i)^n} &= 9.5\end{aligned}$$

Finding,

$$i = 0.063 \rightarrow 6.3\%$$

Thus we have our solution; found either from direct solution with our calculator or from a few simple look-ups in the tables provided by the *FE Reference Handbook*. By comparison, the remodeled home would yield a higher rate-of-return interest after 15 years (6.3% > 4.9%).

**23. THE CORRECT ANSWER IS: A**

2,667 units

---

Refer to the Breakeven Analysis section in the Engineering Economics chapter of the *FE Reference Handbook.*

A product breaks even when the costs to operate are equal to the revenue obtained.

$$\begin{aligned}\text{costs} &= \text{revenue} \\ \$200\,000 + \$50\,x &= \$125\,x \\ x &= 2,667\end{aligned}$$

**24. THE CORRECT ANSWER IS: B**

Present value of periodic repair costs and operating costs

---

An operational facility's lifetime is measured in years and is referred to as the service life of the facility. Life-cycle costing includes the initial cost and the present value of periodic repair and operating costs over the service lifetime.

**25.** **THE CORRECT ANSWER IS: D**

221 mV

---

Refer to Solid-State Electronics and Devices in the Electrical and Computer Engineering chapter of the *FE Reference Handbook.*

The barrier height is also called the junction voltage of the p-n junction. The relationship of the doping impurities on each side of the junction to the junction voltage in the original un-doped case (denoted by subscript 1) is,

$$(V_0)_1 = \frac{kT}{q} \ln \frac{(N_a)_1 (N_d)_1}{n_i^2}$$

The barrier height in the heavily doped sample (denoted by subscript 2) can be written as,

$$(V_0)_2 = \frac{kT}{q} \ln \frac{(N_a)_2 (N_d)_2}{n_i^2}$$

Subtract the two equations to obtain an expression for the difference between the two junction voltages,

$$(V_0)_2 - (V_0)_1 = \frac{kT}{q} \ln \frac{(N_a)_2 (N_d)_2}{(N_a)_1 (N_d)_1}$$

Because the doping of the p-side remains unchanged, then $(N_A)_2 = (N_A)_1$

And $\frac{(N_d)_2}{(N_d)_1} = 5000$. So,

$$\begin{aligned}(V_0)_2 - (V_0)_1 &= 0.026\,\text{V} \times \ln(5000) \\ &= 0.221\,\text{V} \\ &= 221\,\text{mV}\end{aligned}$$

**26.** **THE CORRECT ANSWER IS: D**

5.458 mV

---

Refer to the Solid-State Electronics and Devices section in the Electrical and Computer Engineering chapter of the *FE Reference Handbook.*

The built-in contact potential voltage of a p-n junction is given by,

$$V_o = \frac{kT}{q} \ln \left( \frac{N_a N_d}{n_i^2} \right)$$

The contact potential, $V_o$, is proportional to the temperature, $T$, expressed in Kelvin. The other terms of the expression are equal for both temperature states of the semiconductor.

Convert the temperature from Fahrenheit, °F, to Kelvin, K.

$$°\mathrm{C} = \frac{(°\mathrm{F} - 32)}{1.8}$$
$$K = °\mathrm{C} + 273.15$$
$$K = \frac{(°\mathrm{F} - 32)}{1.8} + 273.15$$

At 90°F, the temperature in Kelvin is 305.37 K

At 190°F, the temperature in Kelvin is 360.92 K

Therefore, the contact potential when the temperature at 190°F is the original contact potential multiplied by the proportional increase in temperature,

$$V_{o_{190°\mathrm{F}}} = 0.030 \times \frac{(360.92)}{(305.37)} = 0.035\,48\,\mathrm{V}$$

Thus the increase in contact potential voltage is the difference between the contact potential voltages at the two temperatures,

$$V_{o_{190°\mathrm{F}}} - V_{o_{90°\mathrm{F}}} = \Delta V_o$$
$$0.035\,48\,\mathrm{V} - 0.03\,\mathrm{V} = 0.005\,458\,\mathrm{V}$$
$$= 5.458\,\mathrm{mV}$$

**27. THE CORRECT ANSWER IS: A**

1.14

---

Refer to the Properties of Materials: Electrical section in the Materials Science/Structure of Matter chapter of the *FE Reference Handbook.*

In the handbook, the definition of $\kappa$ is given as the dielectric constant. The permittivity of a material, $\epsilon$, is also given as the product of the dielectric constant and the permittivity of free space.

Therefore, we can calculate $\kappa$ simply as,

$$\epsilon = \kappa \cdot \epsilon_0$$
$$\frac{\epsilon}{\epsilon_0} = \kappa$$
$$\frac{10.12 \times 10^{-12}\,\mathrm{F\,m^{-1}}}{8.85 \times 10^{-12}\,\mathrm{F\,m^{-1}}} = 1.143 \approx 1.14$$

**28. THE CORRECT ANSWER IS: C**

$0.066\,\mathrm{K}^{-1}$

---

Refer to Resistivity section in the Electrical and Computer Engineering chapter of the *FE Reference Handbook.*

The resistance of a material is given by,

$$R = R_0[1 + \alpha(T - T_0)]$$

According to the problem statement, the new resistance with a temperature increase is,

$$\begin{aligned} 3R_0 &= R_0[1+\alpha(30\,\mathrm{K})] \\ 3 &= 1+\alpha(30\,\mathrm{K}) \\ 2 &= (30\,\mathrm{K})\alpha \\ \alpha &= \frac{2}{30\,\mathrm{K}} = 0.066\,\mathrm{K}^{-1} \end{aligned}$$

**29.** **THE CORRECT ANSWER IS: D**

$$v_{3\,\mathrm{k\Omega}} = 5\,\mathrm{V}, v_{5\,\mathrm{k\Omega}} = 8.335\,\mathrm{V}, v_{7\,\mathrm{k\Omega}} = 11.669\,\mathrm{V}$$

---

Refer to the Kirchhoff's Laws section in the Electrical and Computer Engineering chapter of the *FE Reference Handbook*.

In this circuit, the voltage sources add as you proceed around the loop. To find the voltage across each resistor, write a KVL (loop equation) to discover the current flowing in the loop.

$$5\,\mathrm{V} + 20\,\mathrm{V} = i(3+5+7)\times 10^3\,\mathrm{V}$$

Solve for i,

$$\begin{aligned} i = \frac{25}{15\times 10^3} &= 1.667\times 10^{-3}\,\mathrm{A} \\ &= 1.667\,\mathrm{mA} \end{aligned}$$

Thus, $v_{3\,\mathrm{k\Omega}} = iR_{3\,\mathrm{k\Omega}}$

$$\begin{aligned} v_{3\,\mathrm{k\Omega}} &= (1.667\times 10^{-3})(3\times 10^3)\mathrm{V} = 5\,\mathrm{V} \\ v_{5\,\mathrm{k\Omega}} &= (1.667\times 10^{-3})(5\times 10^3)\mathrm{V} = 8.335\,\mathrm{V} \\ v_{6\,\mathrm{k\Omega}} &= (1.667\times 10^{-3})(7\times 10^3)\mathrm{V} = 11.669\,\mathrm{V} \end{aligned}$$

**30.** **THE CORRECT ANSWER IS: C**

$$-5.77\,\mathrm{V}$$

---

Refer to the Kirchhoff's Laws section in the Electrical and Computer Engineering chapter of the *FE Reference Handbook*.

Define the voltage across the $5\,\mathrm{k\Omega}$ resistor as $V$.

Next, use Kirchoff's Current Law (KCL) to write an equation for the voltage at that node,

$$\frac{V-0}{6\,\mathrm{k\Omega}} + \frac{V-0}{2\,\mathrm{k\Omega}} + \frac{V-0}{5\,\mathrm{k\Omega}} + 5\,\mathrm{mA} = 0$$

Note that resistances in parallel add as,

$$\frac{1}{R_1} + \frac{1}{R_2} + \frac{1}{R_3} = \frac{1}{R_{\mathrm{eq}}}$$

Thus, the summation of the resistances given is,

$$\frac{1}{R_{\text{eq}}} = \frac{1}{6\,\text{k}\Omega} + \frac{1}{2\,\text{k}\Omega} + \frac{1}{5\,\text{k}\Omega} = 0.8667\,\text{k}\Omega^{-1}$$

Note the units are $\text{k}\Omega^{-1}$, so take the reciprocal to find the final $R_{\text{eq}}$, which is,

$$R_{\text{eq}} = 1.15\,\text{k}\Omega$$

Thus

$$\frac{V}{1.15\,\text{k}\Omega} = -5\,\text{mA}$$
$$V = (1.15\,\text{k}\Omega)(-5\,\text{mA}) = -5.77\,\text{V}$$

**31. THE CORRECT ANSWER IS: D**

$$R_{\text{eq}} = 7.7\,\Omega, C_{\text{eq}} = 12\,\mu\text{F}$$

---

Refer to Resistors in Series and Parallel & Capacitors and Inductors in Parallel and Series in the Electrical and Computer Engineering chapter of the *FE Reference Handbook.*

The $5\,\Omega$ and $10\,\Omega$ resistors are in series, and then are in parallel with the $16\,\Omega$ resistor. Combine them to obtain a single resistance, $R_{\text{eq}}$.

$$R_{\text{eq}} = ((5\,\Omega + 10\,\Omega)||16\,\Omega)$$
$$R_{\text{series}} = 5\,\Omega + 10\,\Omega = 15\,\Omega$$

For two resistors in parallel, we can use the shortcut from the handbook,

$$R_{\text{parallel}} = 15\,\Omega||16\,\Omega = \frac{15\,\Omega \cdot 16\,\Omega}{15\,\Omega + 16\,\Omega} = 7.74\,\Omega$$

This means the total equivalent resistance, $R_{\text{eq}} = 7.74\,\Omega$

For the capacitors, the $12\,\mu\text{F}$ and $12\,\mu\text{F}$ are in series, so they can be combined to yield,

$$C_{\text{series}} = \frac{12\,\mu\text{F} \times 12\,\mu\text{F}}{12\,\mu\text{F} + 12\,\mu\text{F}}$$
$$= 6\,\mu\text{F}$$

This will be in parallel with the $6\,\mu\text{F}$ capacitor, therefore,

$$C_{\text{eq}} = 6\,\mu\text{F} + 6\,\mu\text{F} = 12\,\mu\text{F}$$

**32. THE CORRECT ANSWER IS: B**

$$9.17\,\Omega$$

---

Refer to the Resistors in Series and Parallel section in the Electrical Engineering chapter of the *FE Reference Handbook.*

The $10\,\Omega$, $5\,\Omega$, and $10\,\Omega$ resistors to the right most branch of the circuit are in series which sum directly.

The first intermediate series equivalent is therefore

$$R_{\text{eq, series}} = 10\,\Omega + 5\,\Omega + 10\,\Omega = 25\,\Omega$$

This series equivalent is now in parallel with the two $10\,\Omega$ resistors and becomes

$$R_{\text{eq, parallel}} = \frac{1}{\frac{1}{10\,\Omega} + \frac{1}{10\,\Omega} + \frac{1}{25\,\Omega}} = 4.17\,\Omega$$

That parallel resistance can then be taken in series with $5\,\Omega$ yielding the final $R_{\text{eq}}$ across terminals a-b,

$$\begin{aligned} R_{\text{eq}} &= 4.17\,\Omega + 5\,\Omega \\ &= 9.17\,\Omega \end{aligned}$$

**33.** **THE CORRECT ANSWER IS:** $107\,\Omega$

Refer to the Source Equivalents section in the Electrical and Computer Engineering chapter of the *FE Reference Handbook*.

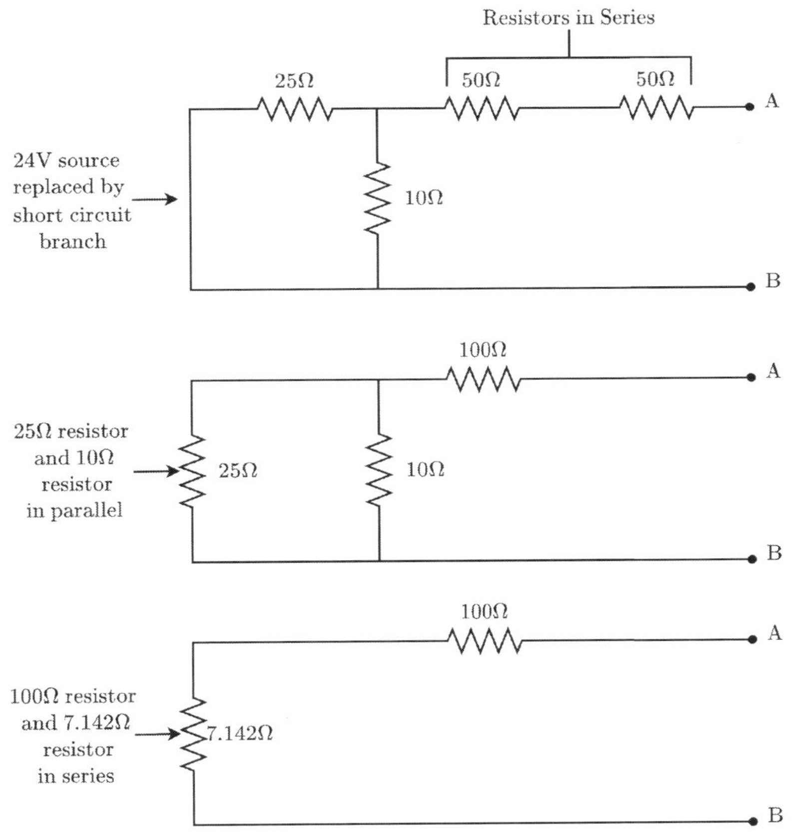


To find the Thevenin equivalent resistance of this circuit, replace all independent voltage sources by a short circuit branch, and calculate the equivalent resistance.

When the 24 V voltage source is replaced by a short-circuit branch, compute the equivalent resistance seen across the terminals A-B.

The two $50\,\Omega$ resistors are in series, and the $25\,\Omega$ and $10\,\Omega$ resistors are in parallel when the voltage source is replaced by a short circuit. The parallel combination of the $25\,\Omega$ and $10\,\Omega$ resistors are then in series with the result of the $50\,\Omega$ resistor series combinations.

Mathematically, this is

$$\begin{aligned} R_{Th} &= 50\,\Omega + 50\,\Omega + (25\,\Omega || 10\,\Omega) \\ &= 100\,\Omega + 7.142\,\Omega \\ &= 107.1\,\Omega \\ &\approx 107\,\Omega \end{aligned}$$

**34. THE CORRECT ANSWER IS: A**

400 V

---

Refer to the Kirchhoff's Laws section in the Electrical and Computer Engineering chapter of the *FE Reference Handbook.*

Write a KCL equation at Node A,

$$-2\,\text{A} + \frac{V_A}{60\,\Omega} - 0.06V_x + \frac{V_A}{120\,\Omega} = 0$$

Rearranged as,

$$\begin{aligned} \left(\frac{1}{60\,\Omega} + \frac{1}{120\,\Omega}\right) V_A &= 2\,\text{A} + 0.06V_x \\ \frac{1}{40\,\Omega} V_A &= 2\,\text{A} + 0.06V_x \end{aligned}$$

$V_x$ can be found from a voltage divider to $V_A$,

$$V_x = \left(\frac{40\,\Omega}{40\,\Omega + 80\,\Omega}\right) V_A = \frac{V_A}{3\,\Omega}$$

Now substitute in the expression for $V_x$,

$$\frac{V_A}{40\,\Omega} = 2\,\text{A} + 0.06\left(\frac{V_A}{3\,\Omega}\right)$$

Solve for $V_A$,

$$\begin{aligned} 0.025\,\Omega^{-1} V_A &= 2\,\text{A} + 0.020\,\Omega^{-1} V_A \\ 0.005\,\Omega^{-1} V_A &= 2\,\text{A} \end{aligned}$$

$$V_A = \frac{2\,\text{A}}{0.005\,\Omega^{-1}} = 400\,\text{V}$$

**35. THE CORRECT ANSWER IS: D**

$Z_C = -j5\,\Omega;\ Z_R = 15\,\Omega;\ V_S = 3.53\angle 30°\,\text{V}$

---

Refer to the Phasor Transforms of Sinusoids section in the Electrical and Computer Engineering chapter of the *FE Reference Handbook*.

The sinusoidal source applied angular frequency is $\omega = 50\,\mathrm{rad\,s^{-1}}$.

The frequency domain impedance of the capacitor is,

$$Z_C = \frac{1}{j\omega C} = -j\frac{1}{(50\,\mathrm{rad/sec})(0.004\,\mathrm{mF})} = -j5\,\Omega$$

The frequency domain impedance of the resistor is simply the value of the resistor, $Z_R = R = 15\,\Omega$. The frequency domain expression for the voltage in phasor form is the $V_{\mathrm{rms}}$ value of the voltage source multiplied by the phase angle. The max amplitude of the source is 5 V so,

$$\begin{aligned} V_{\mathrm{rms}} &= \frac{5\,\mathrm{V}}{\sqrt{2}} \\ &= 3.53\,\mathrm{V} \end{aligned}$$

Finally,

$$V_S = V_{\mathrm{rms}}\angle\phi = 3.53\angle 30°\mathrm{V}$$

**36. THE CORRECT ANSWER IS: A**

1.41 A

---

Refer to the Effective or RMS Values section in the Electrical and Computer Engineering chapter of the *FE Reference Handbook*.

We first work through the full derivation of the solution as a matter of understanding. However, at the end of this solution we make use of a shortcut, noting that this shortcut is only valid for AC signals with no DC component. Such shortcuts are often critical to working through the FE exam in time.

The effective value of a periodic waveform is given by,

$$X_{\mathrm{eff}} = X_{\mathrm{rms}} = \left[\frac{1}{T}\int_0^T x^2(t)dt\right]^{\frac{1}{2}}$$

Where $x(t)$ is the periodic waveform defined over the period $T$.

The waveform given in the figure is $2\sin\omega t$ where $\omega = \frac{2\pi}{T}$ and $T = 4\,\mathrm{m\,sec}$.

Substituting these values into the given equation yields the expression

$$i_{\mathrm{eff}} = \sqrt{\frac{1}{0.004}\int_0^{0.004} 4\sin^2\left(\frac{2\pi t}{0.004}\right)dt}$$

Use the following trigonometric identities (refer to the *FE Reference Handbook*) to substitute $\sin^2 x$ with a function that can be integrated easily,

$$\begin{aligned} \cos 2x &= \cos^2 x - \sin^2 x \\ \sin^2 x &= \cos^2 x - \cos 2x \\ \sin^2 x &= (1 - \sin^2 x) - \cos 2x \end{aligned}$$

Note that $\sin^2 x + \cos^2 x = 1 \Rightarrow \cos^2 x = 1 - \sin^2 x$

$$2\sin^2 x = 1 - \cos 2x$$
$$\sin^2 x = \frac{1}{2}(1 - \cos 2x)$$

Now substitute this identity for $\sin^2 x$ and integrate,

$$i_{\text{eff}} = \sqrt{\frac{4}{0.004}\left[\int_0^{0.004} \frac{1}{2}dt - \int_0^{0.004} \frac{1}{2}\cos\left(\frac{4\pi}{0.004}t\right)\right] dt}$$
$$i_{\text{eff}} = \left[\sqrt{\frac{4}{0.004}\left[\frac{t}{2} - \frac{0.004}{8\pi}\sin\left(\frac{4\pi}{0.004}t\right)\right]}\right]_0^{0.004}$$

Everything is 0 at $t = 0$, so we must only evaluate at $t = 0.004$,

$$i_{\text{eff}} = \sqrt{\frac{4}{0.004}\frac{0.004}{2} - \frac{0.004}{8\pi}\sin\left(\frac{4\pi}{0.004}(0.004)\right)}$$

Since $\sin(4\pi) = 0$, the expression simplifies to,

$$i_{\text{eff}} = \sqrt{\frac{4}{2}}\text{A} = \sqrt{2}\text{A} = 1.414\,\text{A}$$

Shortcut Note: This answer could also have been obtained by taking the peak rms voltage and dividing it by $\sqrt{2}$ which would have also yielded,

$$i_{\text{eff}} = \frac{2\,\text{A}}{\sqrt{2}} = 1.414\,\text{A}$$

This technique only works on sinusoids with no DC voltage, so both solutions are shown for completeness.

**37.** **THE CORRECT ANSWER IS: D**

$3000\angle 90°\text{V}$

---

Refer to the Phasor Transforms of Sinusoids section in the Electrical and Computer Engineering chapter of the *FE Reference Handbook.*

Use Ohm's Law in impedance form to find an expression for the voltage:

$$\mathbf{V} = \mathbf{IZ}$$
$$\mathbf{V} = (150 + j150)\text{mA} \times (10 + j10)\text{k}\Omega$$

Multiply these together to simplify (note that $\text{mA} \times \text{k}\Omega = \text{V}$),

$$\mathbf{V} = [1500 + j1500 + j1500 - 1500]\text{V}$$
$$\mathbf{V} = 0 + j3000\,\text{V} = j3000\,\text{V}$$

Express this in polar form,

$$\mathbf{V} = \sqrt{a^2 + b^2}\angle \tan^{-1}\frac{b}{a}\ \mathrm{V}$$
$$\mathbf{V} = \sqrt{0^2 + 3000^2}\angle \tan^{-1}\frac{3000}{0}\ \mathrm{V}$$
$$= 3000\angle 90^\circ \mathrm{V}$$

**38.** **THE CORRECT ANSWER IS: D**

The magnitude of the rms value

---

Refer to the Complex Power section in the Electrical and Computer Engineering chapter of the *FE Reference Handbook*.

In a phasor diagram, the length of the phasor represents the magnitude of the rms value.

**39.** **THE CORRECT ANSWER IS: A**

$563\angle 27.3^\circ \Omega$

---

Refer to the Resistors in Series and Parallel section in the Electrical and Computer Engineering chapter of the *FE Reference Handbook*.

Remember that impedance of an inductor is $Z_L = j\omega L = jX_L$, and the impedance of a capacitor is $Z_C = \frac{1}{j\omega C}$ or $-j\frac{1}{\omega C} = -jX_C$. Rationalizing the denominator of the capacitor impedance is done so the impedances can add together directly with the $j$ terms.

Therefore, the series impedance is given as,

$$\mathbf{Z} = -jX_C + jX_L + R$$
$$\mathbf{Z} = -j\left(\frac{1}{\omega(120\times 10^{-12}\,\mathrm{F})}\right) + j\omega(15\times 10^{-3}\,\mathrm{H}) + 500\,\Omega$$

If the frequency is 120 kHz, then

$$\omega = 2\pi f = 754\times 10^3\,\mathrm{rad\,s^{-1}}$$

Substituting in the value of $\omega$,

$$\mathbf{Z} = -j\left(\frac{1}{(754\times 10^3\,\mathrm{rad\,s^{-1}})(120\times 10^{-12}\,\mathrm{F})}\right) + j(754\times 10^3\,\mathrm{rad\,s^{-1}}\cdot 15\times 10^{-3}\,\mathrm{H}) + 500\,\Omega$$
$$\mathbf{Z} = -j(1.1052\times 10^4)\Omega + j(1.131\times 10^4)\Omega + 500\,\Omega$$
$$\mathbf{Z} = 500\,\Omega - j258\,\Omega$$

In polar form,

$$\mathbf{Z} = (\sqrt{(500\,\Omega)^2 + (258\,\Omega)^2})(\tan^{-1}\frac{-258\,\Omega}{500\,\Omega})$$
$$\mathbf{Z} = 563\angle 27.3^\circ \Omega$$

**40.** **THE CORRECT ANSWER IS: C**

$0.1 + j0.1\,℧$

---

Refer to the Phasor Transforms of Sinusoids section in the Electrical and Computer Engineering chapter of the *FE Reference Handbook.*

First calculate the impedance. Remember that impedance of a pure resistor is simply $R$, and capacitor impedance is $\frac{1}{j\omega C}$.

The series impedance combination is,

$$Z = R + \frac{1}{j\omega C}$$

Add the two terms together to simplify the sum to obtain,

$$Z = \frac{j\omega CR + 1}{j\omega C}$$

The admittance is defined as the reciprocal of the impedance,

$$Y = \frac{1}{Z} = \frac{j\omega C}{j\omega CR + 1}$$

Now rationalize the denominator of the equation. This is done by multiplying the denominator by its complex conjugate, and this is done to eliminate all imaginary components in the denominator so that the numerator can be split into clear real and imaginary parts,

$$Y = \frac{j\omega C}{1 + j\omega CR} \cdot \frac{1 - j\omega CR}{1 - j\omega CR}$$
$$Y = \frac{j\omega C + \omega^2 C^2 R}{1 + \omega^2 C^2 R^2}$$

Now split the fraction to obtain real and imaginary components of the admittance $Y$,

$$Y = \frac{\omega^2 C^2 R}{1 + \omega^2 C^2 R^2} + j\frac{\omega C}{1 + \omega^2 C^2 R^2}$$

Now, with $\omega = 2000\,\mathrm{rad\,s^{-1}}$,

$$R = 5\,\Omega$$
$$C = 100\,\mu\mathrm{F} = 100 \times 10^{-6}\,\mathrm{F} = 10 \times 10^{-4}\,\mathrm{F}$$

Substitute in the given values,

$$Y = \frac{(2000\,\mathrm{rad/sec})^2(1 \times 10^{-4}\,\mathrm{F})^2 \cdot 5\,\Omega}{1 + (2000\,\mathrm{rad/sec})^2(1 \times 10^{-4}\,\mathrm{F})^2 \cdot (5\,\Omega)^2} + j\frac{(2000\,\mathrm{rad/sec})(1 \times 10^{-4}\,\mathrm{F})}{1 + (2000\,\mathrm{rad/sec})^2(1 \times 10^{-4}\,\mathrm{F})^2 \cdot (5\,\Omega)^2}$$
$$Y = 0.1 + j0.1\,℧$$

**41.** **THE CORRECT ANSWER IS: A**

$877.9\,\mathrm{V}$

---

Refer to the RC and RL Transients section in the Electrical and Computer Engineering chapter of the *FE Reference Handbook.*

An exponentially decaying voltage is equivalent to an RC circuit. Thus, we can use this equation in the *FE Reference Handbook*,

$$v_C(t) = v_C(0)e^{-\frac{t}{RC}} + V(1 - e^{-\frac{t}{RC}})$$

There is no voltage source $V$, so the equation reduces to,

$$v_C(t) = v_C(0)e^{-\frac{t}{RC}}$$

The time constant is RC, which is 400 ms, and the initial value, $v_C(0)$ is 1.2 kV. Substitute in the values to yield,

$$v(t) = (1200)e^{-\frac{t}{400\,\text{ms}}}\,\text{volt}$$

At $t = 125\,\text{ms}$,

$$v(125\,\text{ms}) = 1200e^{-\frac{125\,\text{ms}}{400\,\text{ms}}}\,\text{V}$$
$$v(125\,\text{ms}) = 877.9\,\text{V}$$

**42. THE CORRECT ANSWER IS: C**

1 µF

---

Refer to the Resonance section in the Electrical and Computer Engineering chapter of the *FE Reference Handbook.*

The equation for resonant frequency is,

$$\omega = \frac{1}{\sqrt{LC}}$$

The supply frequency is given in Hz. $\omega = 2\pi f$, thus the equation should be written as,

$$f = \frac{1}{2\pi\sqrt{LC}}$$

Substituting in the given values,

$$5\,\text{kHz} = \frac{1}{2\pi\sqrt{(0.001\,\text{H})C}}$$

Solving the expression for C yields,

$$C = 1.01 \times 10^{-6}\,\text{F} = 1.01\,\mu\text{F} \approx 1\,\mu\text{F}$$

**43. THE CORRECT ANSWER IS: B**

$$\frac{1}{s}[1 - e^{-s}]$$

---

Refer to the Laplace Transform Pairs section in the Mathematics chapter of the *FE Reference Handbook.*

The function given is $f(t) = u(t) - u(t-1)$.

Apply the Laplace Transform Pairs individually,

$$F(s) = \mathscr{L}[u(t)] - \mathscr{L}[u(t-1)]$$

The Laplace Transform of $u(t)$ is $\frac{1}{s}$.

The Laplace Transform of $u(t-1)$ is equivalent to taking the Laplace Transform of the non-time-shifted function and multiplying it by exponent $e^{-st}$. That is, this Transform Pair applies,

$$f(t-\tau)u(t-\tau) \Rightarrow e^{-st}F(s)$$

The $F(s)$ Laplace transform of the non-time shifted unit step $u(t)$ we already know is $\frac{1}{s}$. Thus,

$$F(s) = \frac{1}{s} - e^{-s}\frac{1}{s}$$

Simplifying yields,

$$F(s) = \frac{1}{s}[1 - e^{-s}]$$

**44. THE CORRECT ANSWER IS: A**

$$\frac{1}{(s+1)^2}$$

---

Refer to the Laplace Transform Pairs section in the Mathematics chapter of the *FE Reference Handbook.*

The function given is $f(t) = te^{-t}u(t)$

Apply the Laplace transform pair from the Table that states,

$$f(t) = te^{-\alpha t}u(t) \Rightarrow F(s) = \frac{1}{(s+\alpha)^2}$$

thus,

$$F(s) = \frac{1}{(s+1)^2}$$

The presence of $u(t)$ causes the function to "turn on" at $t = 0$ and will not affect the Laplace transform product.

To show this, the Laplace Transform can also be computed analytically by,

$$\begin{aligned}
(s) &= \int_0^\infty te^{-t}e^{-st}dt \\
&= \int_0^\infty te^{-(s+1)}dt \\
&= -\frac{e^{-(s+1)t}}{(s+1)^2}[-(s+1)t - 1]_0^\infty \\
&= \frac{1}{(s+1)^2}
\end{aligned}$$

Which is the same as the Laplace Transform pair table.

**45.** **THE CORRECT ANSWER IS: B**

$$\frac{2}{3s+2}$$

---

Refer to the Control Systems section in the Instrumentation, Measurement, and Control chapter of the *FE Reference Handbook.*

The transfer function is defined as the ratio of the output to the input in the Laplace domain. Begin by writing a nodal equation of the voltage output.

$$0 = \frac{V_{out} - V_{in}}{R} + C\frac{dV_{out}}{dt}$$

Convert from time-domain to frequency domain by Laplace Transform

$$0 = \frac{V_{out} - V_{in}}{R} + CsV_{out}$$
$$0 = V_{out} - V_{in} + CRsV_{out}$$
$$T(s) = \frac{V_{out}}{V_{in}} = \frac{1}{1 + CRs}$$

Substitute in the numerical values.

$$T(s) = \frac{1}{1 + (6 \times 10^{-6}\text{F})(\frac{1}{4} \times 10^{6}\ )s}$$
$$T(s) = \frac{1}{1 + 1.5s} = \frac{2}{3s+2}$$

**46.** **THE CORRECT ANSWER IS: C**

25 µs

---

Refer to the Sampling section in the Instrumentation, Measurement, and Control chapter of the *FE Reference Handbook.*

The sampling time interval is the inverse of the sampling frequency.

$$T = \frac{1}{\text{sampling frequency}}$$

We are given that we are sampling at 4 times the Nyquist rate.

$$T = \frac{1}{4 \cdot \text{Nyquist rate}}$$

The Nyquist rate is twice the bandwidth of the highest frequency of the system,

$$\begin{aligned} T &= \frac{1}{4 \cdot (2f_m)} \\ &= \frac{1}{4 \cdot (2)(5\,\text{kHz})} \\ &= \frac{1}{40 \times 10^3\,\text{Hz}} \\ &= 25\,\mu\text{s} \end{aligned}$$

**47.** **THE CORRECT ANSWER IS: A**

40 kHz

---

Refer to the Sampling section in the Instrumentation, Measurement, and Control chapter of the *FE Reference Handbook.*

The Nyquist sampling rate must be greater than twice that of the highest frequency present in the signal, 20 kHz. Thus the threshold for the Nyquist sampling rate is,

$$f_s = 2(f_n) = 2(20\,\text{kHz}) = 40\,\text{kHz}$$

**48.** **THE CORRECT ANSWER IS: B**

60 Hz

---

Refer to the Analog Filter Circuits & Operational Amplifiers sections in the Electrical and Computer Engineering chapter of the *FE Reference Handbook.*

Cutoff frequency is defined by the point by which the output signal reduces to $\frac{1}{\sqrt{2}}$ of the input. This is equal to a 3 dB drop in the output or half power drop. This is shown on the frequency plots in the Analog Filters section of the *FE Reference Handbook.*

Because $V_{out}$ is fed back into the negative terminal of the op-amp, then we can assume from ideal op-amp conditions that the voltage at the positive terminal of the op-amp is equal to $V_{out}$.

This means the circuit reduces to that shown below. The following is a full derivation of the equation for cutoff frequency from the given parameters.

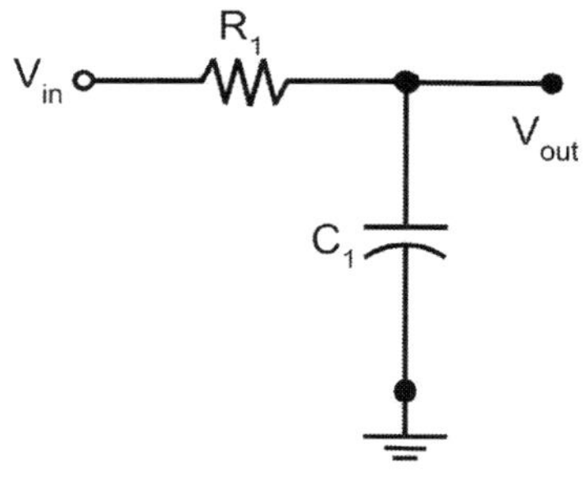


We can now analyze the output voltage circuit as a voltage divider across the capacitor, $C_1$. Note, we will analyze the circuit in terms of the impedances so that the expression can be evaluated in the frequency domain. However, after the full derivation we show a helpful shortcut that can be used on this problem. Such shortcuts are often critical to working through the FE exam in time so when analyzing a problem it is often first useful to think of quick means to the solution.

The impedance of a capacitor is $-jX_C$, where $X_C = \frac{1}{\omega C}$,

$$V_{out} = \frac{-jX_C}{R - jX_C} V_{in}$$

By the definition of the cutoff frequency, the cutoff frequency occurs when the output/input voltage value ratio is $\frac{1}{\sqrt{2}}$, thus,

$$\frac{-jX_C}{R - jX_C} = \frac{1}{\sqrt{2}}$$

Take the magnitude of the expression to yield,

$$\frac{X_C}{\sqrt{R^2 + X_C^2}} = \frac{1}{\sqrt{2}}$$

Factor out $X_C$ from the expression to obtain,

$$\frac{1}{\sqrt{\frac{R^2}{X_C^2} + 1}} = \frac{1}{\sqrt{2}}$$

This implies that,

$$\begin{aligned} \frac{R^2}{X_C^2} &= 1 \\ \frac{R}{X_C} &= 1 \\ R &= X_C \\ R &= \frac{1}{\omega C} \end{aligned}$$

Note that $\omega = 2\pi f$. This value of $f$ is at the cutoff frequency, therefore,

$$f_c = \frac{1}{2\pi RC}$$

Finally, we may substitute in the given values to obtain,

$$\begin{aligned} f_c &= \frac{1}{(2\pi)(10 \times 10^3\,\Omega)(270 \times 10^{-9}\,\mathrm{F})} \\ &= 58.94\,\mathrm{Hz} \\ &\approx 60\,\mathrm{Hz} \end{aligned}$$

Note, the shortcut to this solution involves recognizing the ideal op-amp condition of $V_{out}$ fed back to the negative terminal. Then, the equation for $\omega_c$ could have been used directly from the First-Order Low Pass Filter table in the *FE Reference Handbook*, where,

$$\omega_c = \frac{1}{R_P C} \Rightarrow \frac{1}{R_1 C_1}$$

Recognizing that $R_2$ is very large, so $R_P \approx R_1$.

Substituting in $\omega = 2\pi f$

$$f_c = \frac{1}{2\pi R_1 C_1}$$

$$f_c = \frac{1}{(2\pi)(10 \times 10^3\,\Omega)(270 \times 10^{-9}\,\text{F})}$$

$$= 58.94\,\text{Hz}$$

$$\approx 60\,\text{Hz}$$

**49.** **THE CORRECT ANSWER IS: A**

High-Pass filter

---

Refer to the Analog Filter Circuits section in the Electrical and Computer Engineering chapter of the *FE Reference Handbook*.

Compute the magnitude of the transfer function,

$$|H(j\omega)| = \frac{|j\omega + 0.3|}{|j\omega + 5|}$$

$$= \left(\frac{\omega^2 + 0.09}{\omega^2 + 25}\right)^{\frac{1}{2}}$$

Note the magnitude of $a + bi = \sqrt{a^2 + b^2}$.

Note that $\omega = 2\pi f$ so determine the value of the magnitude of the transfer function as frequency, $f$, increases from $f = 0$. Because the denominator is greater than the numerator, and there is only an $\omega^2$ term in both numerator and denominator, then as $f$ increases, the magnitude of the transfer function increases. This can also be checked by taking the limit of the magnitude as $f \to 0$ and as $f \to \infty$.

When $f = 0$, the magnitude of the transfer function is $\sqrt{\frac{0+0.09}{0+25}} = 0.06$.

When $f \to \infty$ the magnitude of the transfer function approaches the quotient of the coefficients of $\omega^2$, thus, $|H(j\omega)| = 1$ at $\infty$.

This means the filter permits higher frequency signals to pass with a larger magnitude over lower frequency signals, so this is most nearly classified as a high-pass filter. See the illustration shown from the *FE Reference Handbook*'s section on Analog Filter Circuits.

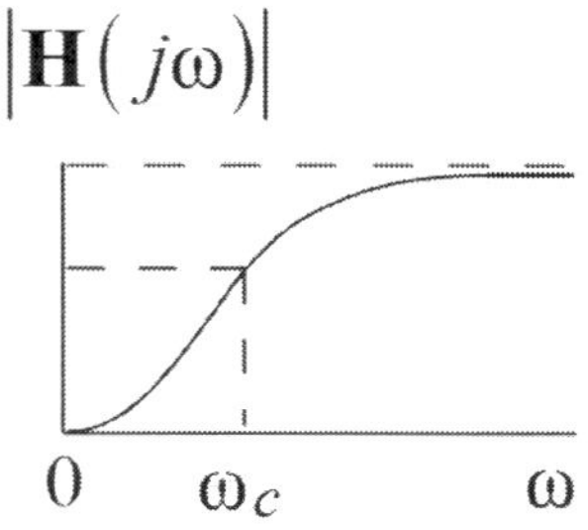


$$|\mathbf{H}(j\omega_c)| = \frac{1}{\sqrt{2}}|\mathbf{H}(j\infty)|$$

**Frequency Response**

**50.** **THE CORRECT ANSWER IS: B**

$$H(z) = \frac{2z + 0.05}{z^2 - 2z + 3}$$

---

Refer to the z-Transforms section in the Electrical and Computer Engineering chapter of the *FE Reference Handbook.*

Take the z-transform of both sides of the equation to yield,

$$Y(z) = Y(z)[2z^{-1} - 3z^{-2}] + X(z)[2z^{-1} + 0.05z^{-2}]$$

The transfer function is ratio of the output to the input,

$$H(z) = \frac{Y(z)}{X(z)}$$

To get $\frac{Y(z)}{X(z)}$, collect the terms to each side of the equation,

$$Y(z) - Y(z)[2z^{-1} - 3z^{-2}] = X(z)[2z^{-1} + 0.05z^{-2}]$$

Factor out $Y(z)$,

$$Y(z)(1 - [2z^{-1} - 3z^{-2}]) = X(z)[2z^{-1} + 0.05z^{-2}]$$

Divide both sides by $X(z)$ and the coefficient of $Y(z)$ to yield $H(z)$,

$$H(z) = \frac{Y(z)}{X(z)} = \frac{2z^{-1} + 0.05z^{-2}}{1 - 2z^{-1} + 3z^{-2}}$$
$$H(z) = \frac{2z + 0.05}{z^2 - 2z + 3}$$

**51.** **THE CORRECT ANSWER IS: A**

$$1, -1.632, 3.399$$

---

Refer to the Numerical Methods and Difference Equations section in the Mathematics chapter of the *FE Reference Handbook.*

Begin by isolating the terms of $y$ to the expression,

$$y[k+1] = -2y[k] + f[k+1]$$

Next, set $k = -1$, and then substitute in the initial condition $y[-1]$ to yield the first term,

$$y[0] = 0 + 1 = 1$$

Set $k = 0$, and use the solution of the previous term to find the next value in the sequence (note: $f(1) = e^{-1} = \frac{1}{e}$),

$$y[1] = -2(1) + \frac{1}{e} = -1.632$$

Set $k = 1$, and use the solution of the previous term to find the next value in the sequence (note: $y[1] = -2 + \frac{1}{e}$ and $f[2] = \frac{1}{e^2}$),

$$y[2] = -2(-2 + \frac{1}{e}) + \frac{1}{e^2} = 3.399$$

**52.** **THE CORRECT ANSWER IS: D**

$$-1.18\,\text{V}$$

---

Refer to the Junction Field Effect Transistors (JFETs) section in the Electrical and Computer Engineering chapter of the *FE Reference Handbook.*

The equation that contains a term for $V_{GS}$ is,

$$i_D = I_{DSS}\left(1 - \frac{V_{GS}}{V_p}\right)^2$$

$$2 = 12\left(1 + \frac{V_{GS}}{2}\right)^2$$

$$\pm 0.408 = \left(1 + \frac{V_{GS}}{2}\right)$$

Therefore the two solutions are

$$V_{GS} = -1.183\,\text{V}, -2.816\,\text{V}$$

However the JFET has a pinch-off voltage at $-2\,\text{V}$, thus $V_P \geq -2\,\text{V}$, and

$$V_{GS} = -1.183\,\text{V}$$

**53.** **THE CORRECT ANSWER IS: C**

$$n = 2.0\ ;\ I_s = 170 \times 10^{-12}\,\text{A}$$

---

Refer to the Solid-State Electronics and Devices section of the Electrical & Computer Engineering Chapter of the *FE Reference Handbook.*

Use the Shockley Equation for a forward biased diode with the given diode voltages. The contribution of the minus 1 in the equation is negligible so it can be ignored in this problem. At 300 K the nominal thermal voltage is 26 mV.

$$i = I_s e^{\frac{vD}{nV_t}}$$

$$0.2 \times 10^{-3}\,\text{A} = I_s \exp\left(\frac{700\,\text{mV}}{n26\,\text{mV}}\right);\ \text{eq. 1}$$

$$4 \times 10^{-3}\,\text{A} = I_s \exp\left(\frac{850\,\text{mV}}{n26\,\text{mV}}\right);\ \text{eq. 2}$$

Divide eq. 2 by eq. 1

$$20 = e^{150/26n} = e^{5.77/n}$$

$$n = \frac{5.77}{\ln 20} = 1.93 \approx 2$$

Back substitute into eq. 2, and solve for $I_s$

$$I_s = 174 \times 10^{-12}\,\text{A} \approx 170 \times 10^{-12}\,\text{A}$$

**54.** **THE CORRECT ANSWER IS: D**

568 Ω

---

Refer to the Bipolar Junction Transistor (BJT) section in the Electrical and Computer Engineering chapter of the *FE Reference Handbook*.

When the transistor is operating under AC conditions, the capacitors begin to act like short circuits. The $V_{CC}$ value goes to zero. Then the $R_C$ resistor comes in parallel with the $R_L$ resistor. Thus, the AC collector resistance is,

$$r_C = R_C || R_L$$

$$r_C = 568\,\Omega$$

**55.** **THE CORRECT ANSWER IS: B**

$I_C = 1.59\,\text{mA}, V_{CE} = 28.8\,\text{V}$

---

Refer to the Bipolar Junction Transistor (BJT) section in the Electrical and Computer Engineering chapter of the *FE Reference Handbook*.

The Q point of a transistor is also called its operating point in the active region for DC voltages and currents when no input signal is applied

To obtain a Q point of 15 V and 4 mA, write an equation for the base loop of the transistor.

$$V_B = R_B I_B + V_{BE} + (I_B + I_C) R_E$$

Substitute in $I_B = \frac{I_C}{\beta}$, and solve for $I_C$,

$$I_C = \frac{(V_B - V_{BE})\beta}{R_B + R_E(1+\beta)}$$

Substitute in the given values, assume $V_{BE} = 0.7\,\text{V}$,

$$I_C = \frac{(4\,\text{V} - 0.7\,\text{V})(100)}{5\,\text{k}\Omega + 2\,\text{k}\Omega(101)} = 1.59\,\text{mA}$$

Now write an equation for the loop of the collector and emitter branch of the BJT to find $V_{CE}$,

$$\begin{aligned}
V_{CC} &= V_{CE} + (R_C + R_B)I_C \\
V_{CE} &= V_{CC} - (R_C + R_B)I_C \\
V_{CE} &= 40\,\text{V} - (5\,\text{k}\Omega + 2\,\text{k}\Omega)(1.59\,\text{mA}) \\
V_{CE} &= 28.8\,\text{V}
\end{aligned}$$

**56.** **THE CORRECT ANSWER IS: A**

$1000, f = 0$

---

Refer to the Operational Amplifiers section in the Electrical and Computer Engineering chapter of the *FE Reference Handbook.*

Compute the magnitude of the given gain expression. The magnitude of the gain is solved as (by taking the magnitude of the complex term in the denominator),

$$|A| = \frac{1000}{\sqrt{(1)^2 + (\omega)^2}}$$

where $\omega = 2\pi f$, and $f$ is the frequency.

For this gain, the magnitude of $A$ is maximum when $f = 0$, and so it therefore occurs at DC. Thus,

$$|A_{max}| = A_{dc} = 1000$$

**57.** **THE CORRECT ANSWER IS: B**

$-\frac{R_2}{R_1}v_{in}$

---

Refer to the Operational Amplifiers section in the Electrical and Computer Engineering chapter of the *FE Reference Handbook.*

Since the problem statement has not told us otherwise, we can assume the op-amp is ideal. The input to the positive terminal is 0, and $v_a = v_{in}$ so this is an inverting op-amp which has the expression for the output voltage of,

$$v_{out} = -\frac{R_2}{R_1}v_{in}$$

Note, it is also possible to compute the output voltage by writing a nodal equation for the negative terminal, and assuming the ideal op-amp conditions,

$$\frac{0\,\text{V} - v_{in}}{R_1} + \frac{0\,\text{V} - v_{out}}{R_2} + 0\,\text{A} = 0\,\text{A}$$
$$\frac{v_{out}}{R_2} = \frac{-v_{in}}{R_1}$$
$$v_{out} = -\frac{R_2}{R_1}v_{in}$$

And this is the same expression as shown by the reference handbook.

**58.** **THE CORRECT ANSWER IS: C**

0.25

---

Refer to the Measurement Uncertainty section in the Instrumentation, Measurement, and Control chapter of the *FE Reference Handbook*. Use the Kline-McClintock equation to find the uncertainty.

$$w_R = \sqrt{\left(w_1\frac{\partial f}{\partial x_1}\right)^2 + \left(w_2\frac{\partial f}{\partial x_2}\right)^2 + \ldots + \left(w_n\frac{\partial f}{\partial x_n}\right)^2}$$

where

$$\frac{\partial f}{\partial x_1} = 5$$
$$\frac{\partial f}{\partial x_2} = -4x_2 = -4(3) = -12$$

$$w_R = \sqrt{(0.01(5))^2 + (0.02(-12))^2}$$
$$\approx 0.25$$

**59.** **THE CORRECT ANSWER IS: D**

40 V

---

Refer to the Power Conversion section in the Electrical and Computer Engineering chapter of the *FE Reference Handbook.*

The voltage gain of a boost converter circuit is given as,

$$\frac{1}{1-D}$$

Remember that gain is defined as the ratio of the output voltage to the input voltage. Thus,

$$\frac{v_{out}}{v_{in}} = \frac{1}{1-D}$$

$D$ is defined as the duty ratio, or duty cycle. The problem statement gives us that the duty cycle is 50%

on and 50% off. Therefore, $D = 0.50$. This gives the whole expression as,

$$v_{out} = \frac{20\,\text{V}}{1 - 0.50} = 40\,\text{V}$$

**60.** **THE CORRECT ANSWER IS: C**

10.68 kvar

---

Refer to the Complex Power section in the Electrical and Computer Engineering chapter of the *FE Reference Handbook.*

The total power received by the station is the summation of the two required powers. Convert the polar forms of the complex power to rectangular for ease of adding them and for analysis at the end of the problem.

$$\mathbf{S}_1 = 30\angle 40^\circ = 22.98 + j19.28$$
$$\mathbf{S}_2 = 15\angle -35^\circ = 12.29 - j8.6$$

Now add the two complex powers together,

$$\mathbf{S}_{total} = \mathbf{S_1} + \mathbf{S_2}$$
$$\mathbf{S}_{total} = 35.27 + j10.68\,\text{kV A}$$

The reactive part of the complex power is the imaginary component of the expression, thus,

$$\mathbf{S}_{\mathbf{total,reactive}} = 10.68\,\text{kvar}$$

**61.** **THE CORRECT ANSWER IS: C**

0.833

---

Refer to the Complex Power section in the Electrical and Computer Engineering chapter of the *FE Reference Handbook.*

The real power in an AC circuit is defined as,

$$P = V_{\text{rms}} I_{\text{rms}} cos\theta$$

Where $cos\theta$ is the power factor. Thus, to compute the power factor, solve for $cos\theta$, and substitute in the given values,

$$cos\theta = \frac{P}{V_{\text{rms}} I_{\text{rms}}}$$
$$cos\theta = \frac{1000\,\text{W}}{(120\,\text{V}_{\text{rms}})(10\,\text{A}_{\text{rms}})} = 0.833$$

**62.** **THE CORRECT ANSWER IS: D**

11+j6 V A

---

Refer to the Complex Power section in the Electrical and Computer Engineering chapter of the *FE Reference*

*Handbook.*

With an applied voltage source of $50\angle 30°$, then the three passive loads are receiving the complex powers described in the problem statement, respectively.

Therefore,

$$\mathbf{S_1} = 4 + j3\,\text{V A}$$
$$\mathbf{S_2} = 5 - j\,\text{V A}$$
$$\mathbf{S_3} = 2 + j4\,\text{V A}$$

With all the loads connected in parallel, then the total power delivered to the loads is the summation of the three complex powers,

$$\mathbf{S} = (4 + 5 + 2) + j(3 - 1 + 4)\,\text{V A}$$
$$\mathbf{S} = 11 + j6\,\text{V A}$$

Total phasor power is the summation of all the power received by the individual circuit elements, regardless of the circuit arrangement or applied source, and is an example of the principle of superposition.

**63.** **THE CORRECT ANSWER IS: C**

$$\mathbf{V}_{an} = 120.09\cos(377t + 0°)\,\text{V}$$
$$\mathbf{V}_{bn} = 120.09\cos(377t - 120°)\,\text{V}$$
$$\mathbf{V}_{cn} = 120.09\cos(377t + 120°)\,\text{V}$$

---

Refer to the Balanced Three-Phase (3-$\phi$) Systems section in the Electrical and Computer Engineering chapter of the *FE Reference Handbook.*

Note that $\omega = 2\pi f = 2\pi(60\,\text{Hz}) = 377\,\text{rad}\,\text{s}^{-1}$

Note also that the problem states that 208 V is given as a peak voltage. The instantaneous form uses the peak voltage, not the $V_{\text{rms}}$ voltage.

For the wye-connection, the line voltage is $V_l = 208\,\text{V}$. Thus, the phase voltage is,

$$V_P = \frac{V_l}{\sqrt{3}} = \frac{208\,\text{V}}{\sqrt{3}} = 120.09\,\text{V}$$

For the positive sequence then, the phase voltages are,

$$\mathbf{V}_{an} = 120.09\angle 0°\,\text{V} = 120.09\cos(377t + 0°)\,\text{V}$$
$$\mathbf{V}_{bn} = 120.09\angle -120°\,\text{V} = 120.09\cos(377t - 120°)\,\text{V}$$
$$\mathbf{V}_{cn} = 120.09\angle 120°\,\text{V} = 120.09\cos(377t + 120°)\,\text{V}$$

And the line voltages are,

$$\mathbf{V}_{ab} = 208\angle 30^\circ \text{ V} = 208\cos(377t + 30^\circ) \text{ V}$$
$$\mathbf{V}_{bc} = 208\angle -90^\circ \text{ V} = 208\cos(377t - 90^\circ) \text{ V}$$
$$\mathbf{V}_{ca} = 208\angle 150^\circ \text{ V} = 208\cos(377t + 150^\circ) \text{ V}$$

**64.** **THE CORRECT ANSWER IS: B**

$$120\angle -90^\circ \ V_{\text{rms}}$$

---

Refer to the Balanced Three-Phase (3-$\phi$) Systems section in the Electrical and Computer Engineering chapter of the *FE Reference Handbook*.

Since the source is balanced, it follows that the magnitudes of all the phase voltages with respect to neutral are equal,

$$|V_{an}| = |V_{bn}| = |V_{cn}|$$

Because it is given that this is a positive phase system, we know that $\mathbf{V}_{bn}$ lags $\mathbf{V}_{an}$ by 120°, and $\mathbf{V}_{cn}$ lags $\mathbf{V}_{bn}$ by 120°.

Since we know the lag between $\mathbf{V}_{bn}$ and $\mathbf{V}_{an}$ and that the magnitude of the voltages are equal, we can compute $\mathbf{V}_{bn}$ directly from the phasor expression,

$$\mathbf{V}_{bn} = |V_{an}|\angle 30^\circ - 120^\circ = 120\angle -90^\circ \ V_{\text{rms}}$$

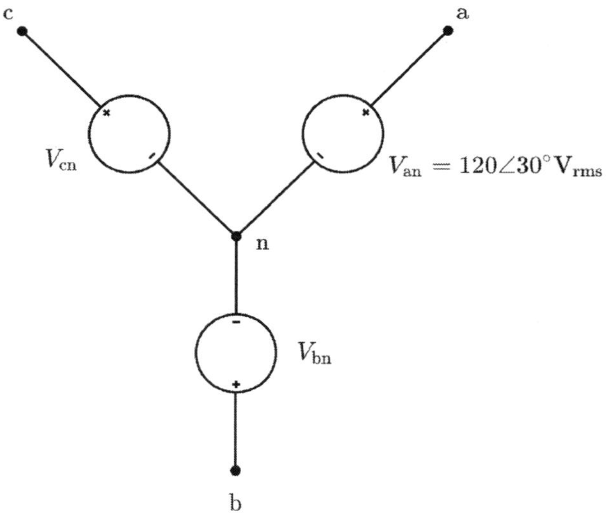


**65.** **THE CORRECT ANSWER IS: D**

75.96 A

---

Refer to the Complex Power section and the Balanced Three-Phase (3-$\phi$) Systems section in the Electrical and Computer Engineering chapter of the *FE Reference Handbook*.

Use the equations,

$$P = V_{\text{rms}} I_{\text{rms}} \cos\theta$$
$$|\mathbf{S}| = \sqrt{3} V_L I_L$$

Solve for the complex power after correction. Let $\mathbf{S}_2$ be the complex power after correction,

$$\mathbf{S}_2 = \frac{50\,\text{kW}}{0.95} = 52.63\,\text{kV A}$$

Now solve for the line current supplied to the load after the power factor correction, $I_L$,

$$\begin{aligned} I_L &= \frac{\mathbf{S}_2}{\sqrt{3} V_L} \\ &= \frac{52.63\,\text{kV A}}{\sqrt{3} \times 400\,\text{V}} \\ &= 75.96\,\text{A} \end{aligned}$$

**66.** **THE CORRECT ANSWER IS: D**

20 V

---

Refer to the Transformers (Ideal) section in the Electrical and Computer Engineering chapter of the *FE Reference Handbook*.

The turns ratio is defined as,

$$a = \frac{N_1}{N_2} = \frac{4}{1}$$

The ratio of the voltages across the primary and secondary are related to the turns ratio by,

$$\frac{V_1}{V_2} = \frac{N_1}{N_2}$$

Thus, the voltage across the secondary is,

$$\begin{aligned} V_2 &= V_1 \frac{N_2}{N_1} \\ &= \frac{120\,\text{V}}{4} \\ &= 30\,\text{V} \end{aligned}$$

In a center-tapped transformer, the center point of the secondary is grounded (see the figure). Thus, half of the secondary voltage appears across the upper half of the transformer and lower half of the secondary windings.

Therefore,

$$\begin{aligned} (V_{upper})_{\text{rms}} &= \frac{1}{2} V_2 \\ &= 15\,\text{V} \end{aligned}$$

The maximum, or peak value, of the rms value is defined from the Effective or RMS Values section in the

Electrical and Computer Engineering chapter of the *FE Reference Handbook.*

$$X_{eff} = X_{\text{rms}} = \frac{X_{max}}{\sqrt{2}}$$

Thus,

$$V_{peak} = \sqrt{2}(V_{upper})_{\text{rms}}$$
$$V_{peak} = 1.414(15\,\text{V}) = 21.21\,\text{V}$$

20 V is most nearly the value of the peak voltage from the available answer options.

**67.** **THE CORRECT ANSWER IS: C**

12 kV

---

Refer to the Turns Ratio section in the Electrical and Computer Engineering chapter of the *FE Reference Handbook.*

The voltage across the load connected at the secondary can be found from applying Ohm's Law,

$$V_s = (I_s)(R_s) = (2\,\text{A})(60\,\Omega) = 120\,\text{V}$$

The primary voltage can be found from the turns ratio expression,

$$\frac{V_p}{V_s} = \frac{100}{1}$$
$$\frac{V_p}{120\,\text{V}} = \frac{100}{1}$$
$$\Rightarrow V_p = 12\,000\,\text{V} = 12\,\text{kV}$$

**68.** **THE CORRECT ANSWER IS: B**

4.6 W

---

Refer to the Rotating Machines (General) section in the Electrical and Computer Engineering chapter of the *FE Reference Handbook.*

We can calculate the mechanical power loss simply from,

$$P = T\omega_m$$
$$P_{\text{mechanical loss}} = (T_{\text{mechanical loss}})(\omega_m)$$
$$P_{\text{mechanical loss}} = (0.0334\,\text{N m})\,(1300\,\text{rpm})\left(\frac{1\,\text{min}}{60\,\text{s}}\right)\left(\frac{2\pi\,\text{rad}}{1\,\text{rev}}\right)$$
$$P_{\text{mechanical loss}} = 4.55\,\text{W}$$

**69.** **THE CORRECT ANSWER IS: B**

The rotational speed of the magnetic field

---

Refer to the AC Machines section in the Electrical and Computer Engineering chapter of the *FE Reference Handbook.*

The synchronous speed, $n_s$, in an AC machine is given by,

$$n_s = \frac{120f}{p}$$

Where $f$ = the line voltage frequency in Hertz, and $p$ = the number of poles of the motor.

The number of poles in a motor is fixed so the synchronous speed is thus principally dependent on the supply voltage frequency. The synchronous speed is the speed of the revolution of the magnetic field in the stator winding of the motor. It is the speed at which the electromotive force is produced by the alternating machine and is measured in revolutions per minute or rpm.

Note that the question asks what the synchronous speed definition is for AC machines in general. In a synchronous machine, the rotor rotates at the same speed as the synchronous speed. However, in an induction motor, the rotor does NOT rotate at the same speed as the synchronous speed and the difference between the rotor's speed, and the synchronous speed is called the slip and is defined as,

$$s = \frac{(n_s - n)}{n_s}$$

Where $n$ is the speed of the rotor measured in rpm

**70.** **THE CORRECT ANSWER IS: D**

$-1130\,\mathrm{V\,m}$

---

Refer to the Electrostatic Fields section in the Electrical and Computer Engineering chapter of the *FE Reference Handbook.*

Gauss's Law for Electricity states that the integral of the electric flux density over a closed surface is equal to the charge enclosed divided by the electric permittivity which can also be expressed as,

$$\frac{Q_{\text{enclosed}}}{\epsilon} = \oiint_S \mathbf{E} \cdot d\mathbf{S}$$

Since we know all the charges enclosed inside the sphere, and that it is a hollow sphere (assumed vacuum), then the flux integral can be computed by simply summing together the contributions of all the individual charge clusters,

$$\begin{aligned}\text{Electric flux} &= \oiint_S \mathbf{E} \cdot d\mathbf{S} \\ &= \frac{Q_1 + Q_2 + Q_3 + Q_4}{\epsilon} \\ &= \frac{+2\,\mathrm{nC} + 3\,\mathrm{nC} - 5\,\mathrm{nC} - 10\,\mathrm{nC}}{8.85 \times 10^{-12}\,\mathrm{F\,m^{-1}}} \\ &= -1129.94\,\mathrm{V\,m}\end{aligned}$$

Note that the units evaluate to Volt-meter. Electric flux can also be expressed as $\mathrm{N\,m^2\,C^{-1}}$

**71. THE CORRECT ANSWER IS: C**

$$\nabla \times \mathbf{B} = -\frac{\partial \mathbf{B}}{\partial t}$$

---

Refer to the Electromagnetic Dynamic Fields section in the Electrical and Computer Engineering chapter of the *FE Reference Handbook*:

The expression, $\nabla \cdot \mathbf{D} = \rho$, is the general form of Gauss' Law for Electricity.

The expression, $\nabla \cdot \mathbf{D} = 0$, is a special case of Gauss' Law for Electricity in a free-space condition.

The expression, $\nabla \times \mathbf{E} = -\frac{\partial \mathbf{B}}{\partial t}$, is the mathematical formulation of Faraday's Law that the cross product of the electric field is equal to the negative rate of change of the magnetic field with respect to time.

The expression, $\nabla \times \mathbf{B} = -\frac{\partial \mathbf{B}}{\partial t}$, is not a correct expression of Ampere's Law, so this is the false statement.

**72. THE CORRECT ANSWER IS: B**

$$1 \leq \text{SWR} < \infty$$

---

Refer to the Lossless Transmission Lines section in the Electrical and Computer Engineering chapter of the *FE Reference Handbook.*

The equation for SWR is given by,

$$\text{SWR} = \frac{1 + |\Gamma|}{1 - |\Gamma|}$$

$\Gamma$ is the load reflection coefficient and is defined as

$$\begin{aligned} \Gamma &= \frac{\mathbf{V}^-}{\mathbf{V}^+} \\ &= \frac{\mathbf{Z}_L - \mathbf{Z}_0}{\mathbf{Z}_L + \mathbf{Z}_0} \end{aligned}$$

For a completely matched line, $\mathbf{Z}_L = \mathbf{Z}_0$, thus, $\Gamma = 0$, and the $SWR = 1$.

For a completely mismatched line, the reflected voltage is equal to the incident voltage. That is, the reflection coefficient $\Gamma$ will be equal to $\pm 1$. When this is the case, the SWR tends towards $\infty$.

Therefore, the SWR will always be in the range,

$$1 \leq \text{SWR} < \infty$$

**73. THE CORRECT ANSWER IS: A**

$$0.83\,\text{A}$$

---

Refer to the Lossless Transmission Lines section in the Electrical and Computer Engineering chapter of the *FE Reference Handbook.*

The incident current $\mathbf{I}^+$ is found from the expression,

$$\mathbf{I}^+ = \frac{\mathbf{V}^+}{\mathbf{Z}_0}$$

Incident voltage is 50 V, and the characteristic resistance is 60 Ω. Therefore,

$$\mathbf{I}^{+} = \frac{50\,\text{V}}{60\,\Omega} = 0.83\,\text{A}$$

**74.** **THE CORRECT ANSWER IS: B**

0.726 s

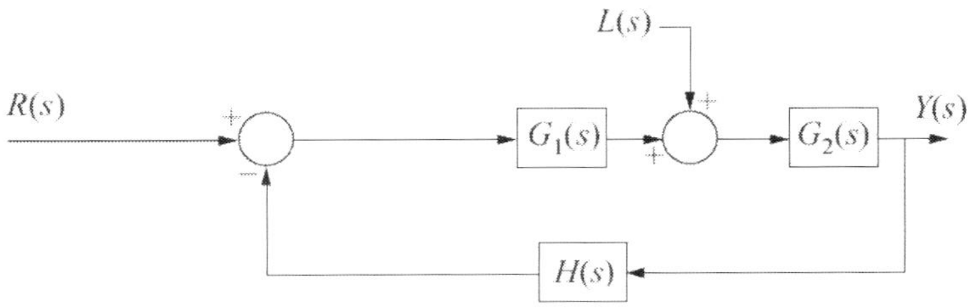


Refer to the Control Systems section in the Instrumentation, Measurement, and Control chapter of the *FE Reference Handbook*.

The controller plant, $G(s)$, is $\frac{25}{s(s+5)}$.

This is a unity feedback controller. It can be deduced that this is a unity feedback controller from looking at the classical model of a negative feedback controller, which is shown in the *FE Reference Handbook* as depicted in the image, and comparing it to our system (noting that $G_2(s) = 1$, $H(s) = 1$ and $L(s) = 0$).

A unity feedback model is also shown explicitly in the section on Control Systems.

Thus, $H(s) = 1$, and the closed loop transfer function, $T(s)$, is $\frac{Y(s)}{R(s)}$ (output over the input) which is,

$$T(s) = \frac{G(s)}{1 + G(s)}$$

Substituting in $G(s)$,

$$\begin{aligned} T(s) &= \frac{\frac{25}{s(s+5)}}{1 + \frac{25}{s(s+5)}} \\ &= \frac{25}{s^2 + 5s + 25} \end{aligned}$$

This is a second-order control system of the form shown in the handbook,

$$\frac{Y(s)}{R(s)} = \frac{K\omega_n^2}{s^2 + 2\zeta\omega_n s + \omega_n^2}$$

Where,

$$\begin{aligned} \omega_n &= \sqrt{25} = 5 \\ 2\zeta\omega_n &= 5 \end{aligned}$$

Substituting $\omega_n$ and solving for $\zeta$ yields,

$$\zeta = 0.5$$

Using the values for $\omega_n$ and $\zeta$, calculate the peak time from,

$$\begin{aligned} t_p &= \frac{\pi}{\omega_n\sqrt{1-\zeta^2}} \\ &= \frac{\pi}{5\sqrt{1-0.5^2}} \\ &= 0.726\,\text{s} \end{aligned}$$

**75. THE CORRECT ANSWER IS: 2**

---

Refer to the Control Systems section in the Instrumentation, Measurement, and Control chapter of the *FE Reference Handbook.*

The poles are the values of $s$ that set the denominator, which is the characteristic equation $D(s)$, equal to zero.

The open-loop characteristic equation is then

$$D(s) = s^2 + 4s + 4$$
$$D(s) = (s+2)^2 = 0$$

This characteristic equation has 2 repeated poles at $s = -2$.

**76. THE CORRECT ANSWER IS: B**

Low-pass filter

---

Refer to the Analog Filter Circuits & Decibels and Bode Plots sections in the Electrical and Computer Engineering chapter of the *FE Reference Handbook.*

Bode plots of the different kinds of first-order filters are shown in the Analog Filters section of the *FE Reference Handbook.* Because this plot represents all frequencies lower than the cut-off frequency being passed, then this bode plot represents a low-pass filter.

**77. THE CORRECT ANSWER IS: B**

$z = 0.9$

---

Refer to the Control Systems section in the Instrumentation, Measurement, and Control chapter of the *FE Reference Handbook.*

The roots of the denominator polynomial are defined as the poles of the transfer function.

$$\begin{aligned} T(z) &= \frac{1}{z^{-2}(1-0.9z^{-1})} \\ &= \frac{1}{\frac{1}{z^2}(1-0.9\frac{1}{z})} \\ &= \frac{z^2}{1-0.9\frac{1}{z}} \\ &= \frac{z^3}{z-0.9} \end{aligned}$$

The pole of the transfer function is $z = 0.9$

**78. THE CORRECT ANSWER IS: A**

1/2

---

Refer to the Control Systems section in the Instrumentation, Measurement, and Control chapter and Second-Order Linear Homogeneous Differential Equations with Constant Coefficients section in the Mathematics chapter of the *FE Reference Handbook.*

The denominator of the transfer function expression is,

$$s^2 + s + 1$$

This has the form of the general second order characteristic equation,

$$s^2 + 2\zeta\omega_n s + \omega_n^2$$

Thus, the undamped natural frequency is,

$$\sqrt{\omega_n} = \sqrt{1} = 1\,\text{rad}\,\text{s}^{-1}$$

Now that $\omega_n$ is known, we can compute the damping ratio, $\zeta$,

$$\begin{aligned} 2\zeta\omega_n &= 1 \\ \zeta &= 1/2 \end{aligned}$$

**79. THE CORRECT ANSWER IS: A**

1.6 s

---

Refer to the Control Systems section in the Instrumentation, Measurement, and Control chapter of the *FE Reference Handbook.*

The controller plant, $G(s)$, is $\frac{25}{s(s+5)}$. This is a unity feedback controller, thus, $H(s) = 1$, and the closed loop transfer function, $T(s)$ is,

$$T(s) = \frac{G(s)}{1 + G(s)}$$

Substituting in $G(s)$,

$$T(s) = \frac{\frac{25}{s(s+5)}}{1+\frac{25}{s(s+5)}}$$
$$= \frac{25}{s^2+5s+25}$$

This is a second-order control system where,

$$\omega_n = \sqrt{25} = 5$$
$$2\zeta\omega_n = 5$$

Substituting $\omega_n$ and solving for $\zeta$ yields,

$$\zeta = 0.5$$

Using the value for $\omega_n$ and $\zeta$, calculate the settling time from,

$$T_s = \frac{4}{\zeta\omega_n}$$
$$= \frac{4}{(0.5)(5)}$$
$$= 1.6\,\text{s}$$

**80. THE CORRECT ANSWER IS: C**

10

---

Refer to the Pulse-Code Modulation (PCM) section in the Electrical and Computer Engineering chapter of the *FE Reference Handbook.*

The number of quantization levels for a binary word of length n-bits to be encoded for a given sample is defined by,

$$q = 2^n$$

The quantization levels are given at 1024. Therefore,

$$1024 = 2^n$$

Take the logarithm of both sides,

$$\log 1024 = n \cdot \log 2$$

Solve for $n$,

$$\frac{\log 1024}{\log 2} = n = 10$$

Therefore, 10 binary bits, or pulses, are required to encode the signal.

**81.** **THE CORRECT ANSWER IS: B**

1200 Hz

---

Refer to the Pulse-Code Modulation (PCM) section in the Electrical and Computer Engineering of the *FE Reference Handbook.*

This is a pulse-code modulated signal. Channel capacity, $C$, is defined as at least $2nW$ binary pulses per second (or bits per second).

Thus, we can use the expression

$$C = 2nW = 2W \log_2 q$$

Where $W$ is the bandwidth and $q$ is the quantization level. We know the signal is encoded into a 4-bit word, so $\log_2 q = 4$.

We can substitute the values as,

$$9600 = 2W \times 4$$

And solve for the bandwidth,

$$\begin{aligned} W &= \frac{9600\,\text{bps}}{8} \\ &= 1200\,\text{Hz} \end{aligned}$$

**82.** **THE CORRECT ANSWER IS: D**

$$\frac{2}{1 + 2j\omega}$$

---

Refer to the Fourier Transform section in the Mathematics chapter of the *FE Reference Handbook.*

We show two ways of solving this problem. First we work through the full derivation of the solution as a matter of understanding. However, at the end of this solution we make use of a shortcut; such shortcuts are often critical to working through the FE exam in time.

The full derivation is shown below for completeness and as an example of what to do if the function cannot be found in the Fourier Transform table.

The definition of the Fourier transform is:

$$F(\omega) = \int_{-\infty}^{\infty} f(t)e^{-j\omega t}dt$$

Thus, substituting in the given values yields,

$$F(\omega) = \int_{-\infty}^{\infty} e^{-t/2}e^{-j\omega t}u(t)dt$$

The unit step function is u(t) and has the following definitions (refer to the Digital Signal Processing section in the *FE Reference Handbook*),

$$u(t) = \begin{cases} 0 & ,t < 0 \\ 1 & ,t \geq 0 \end{cases}$$

It has the effect of "turning on" the functions it is multiplied by at t = 0. Thus the integral can be simplified to,

$$F(\omega) = \int_0^{\infty} e^{-t/2} e^{-j\omega t} dt$$
$$F(\omega) = \int_0^{\infty} e^{-(\frac{1}{2}+j\omega)t} dt = \int_0^{\infty} e^{-(\frac{1+2j\omega}{2})t} dt$$

Evaluating the integral,

$$F(\omega) = -\frac{2}{1+2j\omega} e^{-(\frac{1+2j\omega}{2})t} \Big|_0^{\infty}$$
$$F(\omega) = -\frac{2}{1+2j\omega} \frac{1}{e^{\frac{1+2j\omega}{2}\infty}} - \left(-\frac{2}{1+2j\omega} \frac{1}{e^{\frac{1+2j\omega}{2}0}}\right)$$

When t $\rightarrow \infty$, the first term exponential$\rightarrow$ 0. When $t \rightarrow 0$, the second term exponential $\rightarrow 1$, thus the final form of the Fourier transform,

$$F(\omega) = \frac{2}{1+2j\omega}$$

The fastest way to solve this problem is to note that the answer can be found from the Fourier Transform Pairs table in the *FE Reference Handbook* where,

$$e^{-at}u(t) \Leftrightarrow \frac{1}{a+j2\pi f}$$

Where $a = \frac{1}{2}$ in this problem, and $\omega = 2\pi f$ (angular frequency by definition). Plugging in,

$$F(\omega) = \frac{1}{\frac{1}{2}+j\omega} = \frac{2}{1+2j\omega}$$

**83.** **THE CORRECT ANSWER IS: B**

$$8\,\mathrm{sinc}(4f)e^{-4j\pi f}$$

---

Refer to The Fourier Transform and its Inverse & Communication Theory and Concepts in the *FE Reference Handbook.*

The function shown is a rectangular pulse function center on 2, with an amplitude of 2 and a total duration of 4 ms.

Note, a rectangular pulse has the form shown below.

$$\Pi\left(\frac{t}{\tau}\right)$$

This is shown explicitly in the Communication Theory section of the *FE Reference Handbook* and implicitly in the Fourier Transforms section beneath the Fourier Transform Pairs Theorems table.

Substituting in the amplitude, phase shift, and phase duration into the form of a rectangular pulse yields the expression of the repeated product shown as,

$$2\Pi\left(\frac{t-2}{4}\right)$$

Referring to the Fourier Transform Pairs table, the Fourier transform of this function with the time shift exponential term theorem (shown in the Fourier Transform Theorems table) becomes,

$$2[4\text{sinc}(4f)]e^{-j2\pi f(2)} = 8\text{sinc}(4f)e^{-4j\pi f}$$

**84. THE CORRECT ANSWER IS: A**

12.6 kHz

---

The answer to this is not directly searchable in the handbook but it is easy to deduce if one understands the definitions used. A guard band is an unused portion of the communication spectrum to prevent interference between two communication bands. See the figure.

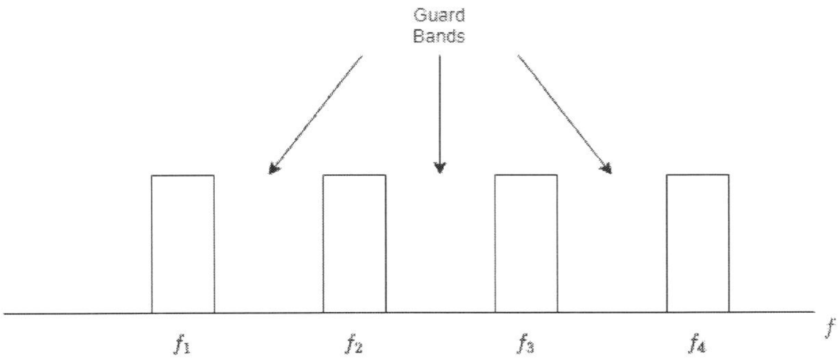


There are 4 devices each requiring 3000 Hz, and there is a 200 Hz guard band between each device, which means 3 guard bands are required to separate the 4 devices. So, the bandwidth required for the channel is simply,

$$\text{Bandwidth} = (3000 \times 4) + (200 \times 3) = 12.6\,\text{kHz}$$

**85. THE CORRECT ANSWER IS: A, D**

00, 11

---

Refer to the Error Coding section in the Electrical and Computer Engineering chapter of the *FE Reference Handbook*.

Since the system uses odd parity, an odd number of '1' bits (including the parity bit) should be received.

There are 3 '1's being received by the stream. So, in order to have odd parity, the leading bits can be either 00 or 11, for a total of 3 '1's being received or 5 '1's being received.

**86. THE CORRECT ANSWER IS: C**

Digital communication can withstand channel noise and distortion within limits

---

A main advantage of digital communications over analog communications is that any distortions present will distort an analog signal. Digital signals are more tolerant of distortions and channel noise.

Digital communications do require repeaters for long transmission distances as accumulated distortion can overwhelm their tolerances.

Digital communications is not immune to entropy in transmission.

Transmitting an analog signal digitally requires encoding the analog signal into digital form.

**87. THE CORRECT ANSWER IS: B**

the process of finding a path on which data can pass from source to destination.

---

Refer to the Computer Networking section in the Electrical and Computer Engineering chapter of the *FE Reference Handbook*. The answer to this question is not directly searchable in the handbook but it can be inferred from this section from the handbook,

"The network layer or Internet layer adds another header normally containing the IP protocol; the main role of the networking layer is finding appropriate routes between end hosts, and forwarding the packets along these routes."

Routing is the process of finding a path on which data can pass from source to destination. Routing is done by a device called routers, which are network layer devices.

**88. THE CORRECT ANSWER IS: C**

To eliminate signal bounce

---

Refer to the Network Topologies section in the Electrical and Computer Engineering chapter of the *FE Reference Handbook*.

In a bus network topology, all stations are connected to a backbone main bus line that runs the entire length of the network. The signals are transmitted along the whole length of the line. If there are no terminators at each end of the line, then the signals will be reflected at the unterminated end and travel back up the line infinitely causing self-interference with other signals propagating on the line. In short, they eliminate signal bounce by absorbing the signal after each transmission and allow new signals to propagate freely on the line by preventing standing waves from developing on the line. Refer to the Lossless Transmission Lines section.

Crosstalk refers to electromagnetic interference that occurs when one communication channel causes distortions in another - typically by capacitive, inductive, or conductive coupling. Terminators can help with *reducing* crosstalk but terminators alone cannot eliminate crosstalk.

**89.** **THE CORRECT ANSWER IS: C**

Domain Name System

---

Refer to the Nmap section in the Electrical and Computer Engineering chapter of the *FE Reference Handbook*.

It is not possible to look up the definition of DNS directly in the *FE Reference Handbook* but the acronym is used in the Nmap section, and it is expected that engineers shall know what this common term means.

DNS stands for "domain name system" and is the hierarchical, decentralized naming system for computers, services, and other resources connected to the Internet or a private network.

**90.** **THE CORRECT ANSWER IS: A**

Data link

---

Refer to the Computer Networking section in the Electrical and Computer Engineering chapter of the *FE Reference Handbook*.

This exists in the Data Link layer as a 33.6 kbps modem uses framing, flow-control, and error correction to connect a user to the switch. It handles node-to-node data transfer.

Note this is not the Physical layer as the modem acts above the actual copper that the data flows on which is the domain of the Physical layer (pin layout, impedance, voltage, signal timing, and frequency).

**91.** **THE CORRECT ANSWER IS: A**

Fragmentation

---

Refer to the Firewall/IDS Evasion and Spoofing section in the Electrical and Computer Engineering chapter of the *FE Reference Handbook*. It is not possible to look up the answer to this question directly in the *FE Reference Handbook*.

Fragmentation is an intrusion detection system (IDS) evasion method. It attacks the computer by sending the attack as multiple split up packets and sent over varying time intervals. In order to stop the attack, the IDS must be able to reassemble the packets and recognize that an attack is occurring.

Denial of service and access attacks are not evasion methods for directly defeating the IDS system as their method rely first on overwhelming the computer defenses.

Pattern matching is a technique for detecting IDS evasion, not evading the IDS.

**92.** **THE CORRECT ANSWER IS: A**

False positive

---

Refer to the Computer Network Security section in the Electrical and Computer Engineering chapter of the *FE Reference Handbook*. It is not possible to directly look up the answer in the *FE Reference Handbook*. But the answer can be deduced from the available choices.

This is a False Positive because it was not an actual malicious attack but resulted in the generation of an alarm, this alarm is referred to as a false positive. A false negative occurs when an actual attack is not reported. True positives occur when real attacks are successfully detected and reported. True negatives happen when no attack occurred, and no alarm was generated.

**93. THE CORRECT ANSWER IS: C**

$$37_{\text{hex}}$$

---

Refer to the Binary Number System section in the Electrical and Computer Engineering chapter of the *FE Reference Handbook.*

The value can be converted by splitting the byte into 2 sets of 4 bit values (called nibbles) and using the binary-to-hexadecimal table in the *FE Reference Handbook.*

$$[0011]_2 = 3_{\text{hex}}$$
$$[0111]_2 = 7_{\text{hex}}$$

Therefore,

$$[00110111]_2 = 37_{\text{hex}}$$

**94. THE CORRECT ANSWER IS: 25**

---

Refer to the Binary Number System section in the Electrical and Computer Engineering chapter of the *FE Reference Handbook.*

The decimal equivalent of a binary number is given by

$$D = a_k 2^k + a_{k-1} s^{k-1} + ... + a_0 + a_{-1} 2^{-1} + ...$$

Thus the number is,

$$\begin{aligned} D &= 2^4 + 2^3 + 0 + 0 + 2^0 \\ &= 16 + 8 + 1 \\ &= 25 \end{aligned}$$

**95. THE CORRECT ANSWER IS: B**

$$A + C$$

---

Refer to the Logic Operations and Boolean Algebra section of the Electrical and Computer Engineering chapter of the *FE Reference Handbook.*

Begin by factoring A,

$$(A + C)A(D + \bar{D}) + AC + C$$

Apply the complement identity $D + \bar{D} = 1$,

$$A(A + C) + AC + C$$

Order the terms so that A can be factored from,

$$A((A + C) + C) + C$$

The terms of C inside the parentheses can be simplified by the identity $C + C = C$,

$$A(A + C) + C$$

Multiply A back through the expression,

$$AA + AC + C$$

Apply identity $AA = A$,

$$A + AC + C$$

Factor out $C$,

$$A + (A + 1)C$$

Apply the identity $A + 1 = 1$,

$$A + C$$

**96.** **THE CORRECT ANSWER IS: C**

NOR or XNOR

---

Refer to the Logic Operations and Boolean Algebra section in the Electrical and Computer Engineering chapter of the *FE Reference Handbook*.

To answer this question, determine the output of each logic gate under the given conditions of all inputs being '1'. Assume inputs A and B.

The output of an AND gate is $AB$. When all its inputs are 0 the output is 0.

The output of an OR gate is $A + B$. When all its inputs are 0 the output is 0.

The output of a NAND gate is $\overline{AB}$. When all of the inputs are 0 the output is 1. This means NAND gates are a possible answer.

The output of a NOR gate is $\overline{A + B}$. When all the inputs are 0 the output is 1. This means NOR gates are a possible answer.

The output of an XOR gate is only 1 for an odd number of 1's at the input. Since all the inputs are 0, then the output is 0.

The output of an XNOR gate is the inverse, or complement, of the XOR gate. Since the XOR gate output is 0 in this situation with all logic inputs being 0, then the output of XNOR gate is 1. This means XNOR gate is a possible answer.

The only answer that contains 2 possible right solutions is NOR or XNOR.

**97.** **THE CORRECT ANSWER IS: C**

$$F = AC + BC$$

---

Refer to the Karnaugh Maps section in the Electrical and Computer Engineering chapter of the *FE Reference Handbook.*

There are multiple ways to solve a Karnaugh map, but the easiest and simplest method is to make as many large groupings as possible.

The largest grouping can be made of all the high values in the lower right of the K-map.

In row 2, column 4, $C$ must be high, and $A$ must be high, yielding $AC$.

In row 2, columns 2 and 3, $B$ must be high, and $C$ must be high. $A$ is a don't care condition because the output is high regardless of the value of A.

This show that the values AC are high and BC are high. In sum-of-products form,

$$F = AC + BC$$

**98.** **THE CORRECT ANSWER IS: B**

$$Q = 1$$
$$\overline{Q} = 0$$

---

Refer to the Flip-Flops section in the Electrical and Computer Engineering chapter of the *FE Reference Handbook.*

See the image below for a timing diagram of $Q$ and $\overline{Q}$. In this flip-flop, $B$ is the clock pulse. The outputs, $Q$ and $\overline{Q}$, respond to the input, $A$, while the CLK pulse, $B$ is high.

When the first CLK pulse occurs, $Q$ toggles to 1 and then back to 0, and $\overline{Q}$ toggles to 0 and then back to 1.

When the second CLK pulse occurs, $Q$ toggles to 1 and $\overline{Q}$ toggles to 0, and they both stay there since the input $A$ does not change before the CLK pulse $B$ goes low.

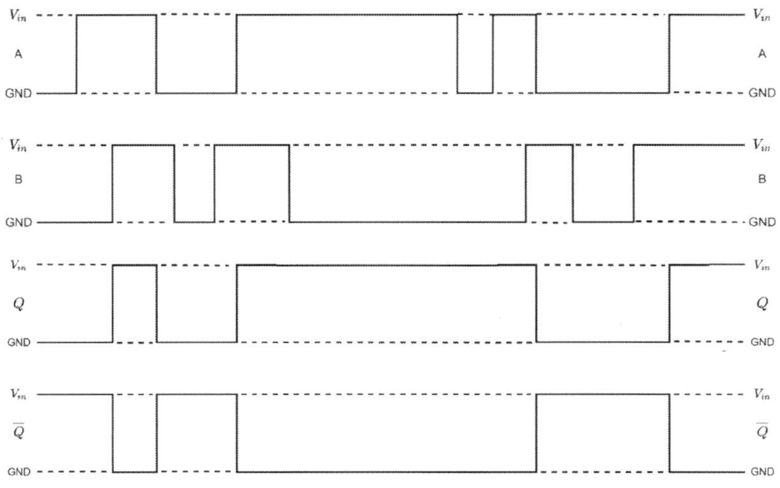

**99.** **THE CORRECT ANSWER IS: D**

$$\bar{A}\bar{B}C + \bar{A}B$$

---

Refer to the Logic Operations and Boolean Algebra section in the Electrical and Computer Engineering chapter of the *FE Reference Handbook.*

The large dots indicate connection points so begin by tracing the paths of each input signal from left to right.

The leftmost AND gate is not connected to output Y.

The second AND gate is not connected to output Y.

The third AND gate has inputs $\bar{A}$, $\bar{B}$, and $C$ connected to its input which is then fed to the output Y. The first term is then,

$$\bar{A}\bar{B}C$$

The fourth AND gate has inputs $\bar{A}$, and $B$ connected to its input which is then fed to the output Y. The second and final term is,

$$\bar{A}B$$

Combining these terms as the output of the OR gate to output Y yields,

$$\bar{A}\bar{B}C + \bar{A}B$$

**100.** **THE CORRECT ANSWER IS: B**

0/0

---

Refer to the Finite State Machine section in the Mathematics chapter of the *FE Reference Handbook.*

Examine the state diagram and look at State B. There are two state inputs possible, 0 and 1. When the input is 0, the output is 0, and the machine transitions to State C. For completeness, when the input is 1, the output is 0, and the machine transitions to State B. Thus, the input/output combination that causes the machine to transition from State B to State C is 0/0.f

**101.** **THE CORRECT ANSWER IS: D**

Output Z will glitch to logic 1 for a moment

---

Refer to the Logic Operations and Boolean Algebra section in the Electrical and Computer Engineering chapter of the *FE Reference Handbook.*

The scenario described is an example of a race condition. That is, the output at $Z$ depends on the time of the arrivals of each input to the AND gate.

Construct a timing diagram to visualize the state transition of the output when the input $A$ changes from 0 to 1 and if the output from the inverter, $\bar{A}$, is delayed by some arbitrary time.

If the change in output of the inverter is delayed by an arbitrary time, then there is a period of time when the logic output of the inverter is 1, and the logic input $A$ is also 1, yielding a total output $Z$ of 1 for a moment. This is incorrect for the desired logic expression and is a 'glitch' in the logic output $Z$.

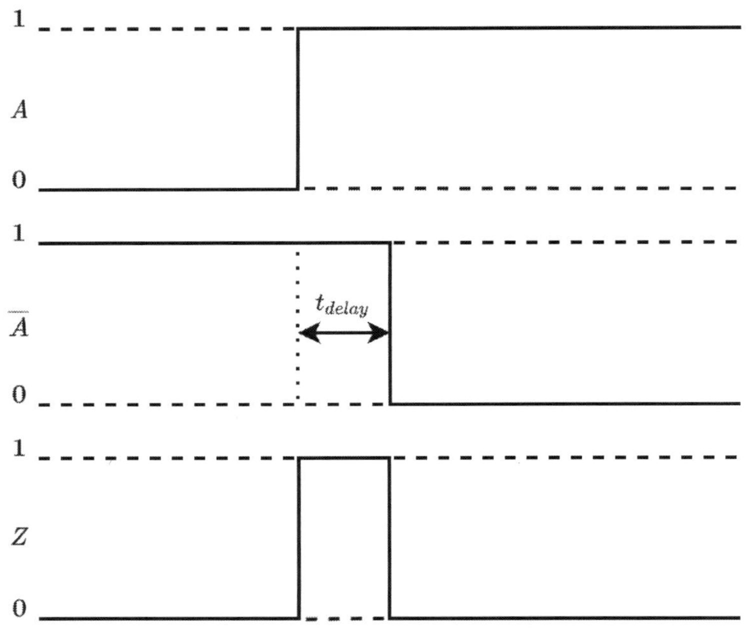


**102.** **THE CORRECT ANSWER IS: C**

Address bus

---

Refer to the Computer Systems section in the Electrical and Computer Engineering chapter of the *FE Reference Handbook*.

The solution to this question is not searchable directly in the *FE Reference Handbook*. However, if one refers to the graphic showing the Microprocessor Architecture - Harvard (reproduced below), one can see that the architecture explicitly depicts a block that calls out the Memory portions of the microprocessor. To store information in Memory, the computer needs a Memory Address. The bus that chooses this location is therefore referred to as the Address bus.

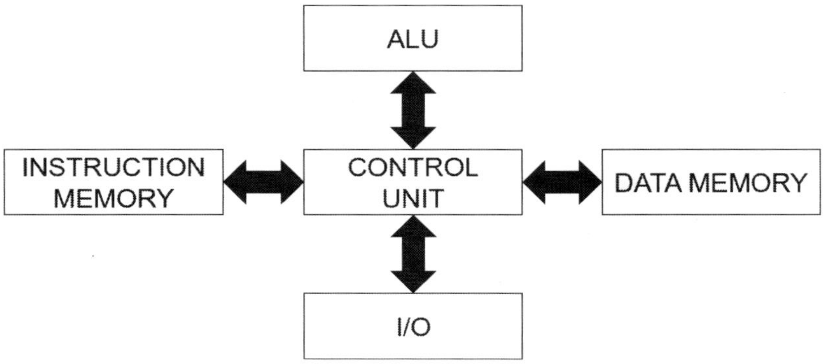


**103.** **THE CORRECT ANSWER IS: C**

Division

---

It is not possible to directly search this in the *FE Reference Handbook.*

In an 8086 microprocessor there are 6 status flags and 3 control flags. They are:

- Sign flag (S)
- Zero flag (Z)
- Auxiliary carry flag (AC)
- Parity flag (P)
- Carry flag (CY)
- Overflow flag (O)

The control flags are:

- Directional flag (D)
- Interrupt flag (I)
- Trap flag (T)

This is an example of a general knowledge question that can be encountered on the exam which is not referenced in the handbook. Of these choices, one may reasonably deduce that Division is not a flag since standard operations in microprocessor bit registers are addition and subtraction which have parity, sign, and carry components.

**104. THE CORRECT ANSWER IS: D**

EPROM can be erased by UV light

---

Refer to the Computer Systems section in the Electrical and Computer Engineering chapter of the *FE Reference Handbook.*

The solution to this question is not searchable directly in the *FE Reference Handbook.* However, EPROM is a type of a ROM. ROM stands for "read only memory". The EP in EPROM stands for "erasable programmable".

EPROMs are programmed by applying a voltage to a pair of transistors which causes electrons to become embedded in an oxide layer separating them, yielding a net negative charge and setting the value of the logic gate to either 0 or 1. They are used to permanently store data as once they are programmed they are not easily erased. However, they can be erased by applying UV light to the transistor layers and releasing the embedded electrons. After this, they may be reprogrammed.

**105. THE CORRECT ANSWER IS: B**

Cache memory, main memory, hard disk, CD-ROM

---

Refer to the Memory/Storage Types section in the Electrical and Computer Engineering chapter of the *FE Reference Handbook.*

Cache memory is the memory closest to the CPU and is the fastest. Main memory is the next closest level up from the CPU in the given list. Hard disk memory is typically installed directly into the computer system and are the next level up from the CPU on the given list. CD-ROMs are read from a low power laser beam which must then be processed through several layers before being interpreted by the CPU which is the slowest method of all on the given list.

**106. THE CORRECT ANSWER IS: A, B, C**

- The path of communication between two components
- Accessing memory
- The method a user employs to interact with the components

---

The microprocessor connects with many internal and external devices to process tasks. This procedure is called Interfacing in a microprocessor.

The microprocessor has I/O interfacing and memory interfacing. The connectivity of input devices (keyboard) and output devices (screen) are examples of microprocessor I/O interfacing.

The microprocessor accessing the memory to read the instruction code and store the data is called memory interface.

**107. THE CORRECT ANSWER IS: A**

$$\mathrm{O}(n \cdot \log n)$$

---

Refer to the Algorithm Efficiency (Big-O) section in the Electrical and Computer Engineering chapter of the *FE Reference Handbook.*

The total time taken will be N times the complexity of adding a single element to the heap. And adding a single element takes $\log n$ time. Therefore, the total time complexity is to $n \cdot \log n$.

**108. THE CORRECT ANSWER IS: B**

Coupling Operation

---

Refer to the Data Structures section in the Electrical and Computer Engineering chapter of the *FE Reference Handbook.*

A skip list is a type of linked list (see the Data Structures section for linked list definition) that allows the data to be accessed by skipping nodes to the location of the desired data. A skip list is constructed with a series of layers with higher level layers being used as an express lane to skip nodes to access data in sub-layers more quickly.

There are three standard operations in skip lists - insertion (to add elements), deletion (to remove elements), and search (to find elements).

**109. THE CORRECT ANSWER IS: A**

Must know the maximum number of nodes possible before creating the trees

---

Refer to the Data Structures section in the Electrical and Computer Engineering chapter of the *FE Reference Handbook*.

The size of an array is static in fixed arrays. In order to create the tree, we need to know the number of nodes in the tree before array declaration. It is the main disadvantage of using fixed arrays to represent trees.

**110.** **THE CORRECT ANSWER IS: B**

0.54

---

Follow the algorithm instructions, and keep repeating the `while` loop until it no longer applies.

**First algorithm iteration**:

- Start: $i = 13.0, j = 4.0$
- Step 0: is $j \leq 6$?...True, $j = 4.0$
- Step 1: $i = i/j \Rightarrow 13.0/4.0 = 3.25 \therefore i = 3.25$
- Step 2: $j = j + 2 \Rightarrow 4.0 + 2 \therefore j = 6.0$

**Second algorithm iteration**:

- Start: $i = 3.25, j = 6.0$
- Step 0: is $j \leq 6$?...True, $j = 6.0$
- Step 1: $i = i/j \Rightarrow 3.25/6.0 = 0.54 \therefore i = 0.54$
- Step 2: $j = j + 2 \Rightarrow 6.0 + 2 \therefore j = 8.0$

**Third algorithm iteration**:

- Start: $i = 0.54, j = 8.0$
- Step 0: is $j \leq 6$?...False, $j = 8.0$
- Step 1: end WHILE loop
- Step 2: Print $i$... $\therefore i = 0.54$

Algorithm and logic problems, such as this one, are part of the FE Exam topics/specifications list. There's no equation in the *FE Reference Handbook*, or any handbook, to solve algorithms. Solve any algorithm by strictly following the instructions the algorithm is presenting to you.

# Additional Practice

At PrepFE, we strongly encourage preparing for the FE exam by practicing as many problems as you can. Active and engaged learning like this is the best way to prepare for The FE exam. We provide hundreds of problems with detailed and thorough solutions and explanations, and you can practice with several exam variations, such as timed exams or with a focus on specific categories. Problems at PrepFE.com are different from those in this practice exam book. Visit `https://prepfe.com` to learn more!

# Scratch Paper & Answer Sheet

Use the following pages for your extra notes while working through these problems (this exam is not a work of art, write all over every page as you work through the exam!). The very last page in the book is an answer sheet you can quickly flip back to and fill out answers as you progress through the practice exam. You may also use this answer sheet as your reference if you choose to upload your results to our performance analytics website, at `https://prepfe.com/performance` or by following the QR code below.

| | | | |
|---|---|---|---|
| 1. | 29. | 57. | 85. |
| 2. | 30. | 58. | 86. |
| 3. | 31. | 59. | 87. |
| 4. | 32. | 60. | 88. |
| 5. | 33. | 61. | 89. |
| 6. | 34. | 62. | 90. |
| 7. | 35. | 63. | 91. |
| 8. | 36. | 64. | 92. |
| 9. | 37. | 65. | 93. |
| 10. | 38. | 66. | 94. |
| 11. | 39. | 67. | 95. |
| 12. | 40. | 68. | 96. |
| 13. | 41. | 69. | 97. |
| 14. | 42. | 70. | 98. |
| 15. | 43. | 71. | 99. |
| 16. | 44. | 72. | 100. |
| 17. | 45. | 73. | 101. |
| 18. | 46. | 74. | 102. |
| 19. | 47. | 75. | 103. |
| 20. | 48. | 76. | 104. |
| 21. | 49. | 77. | 105. |
| 22. | 50. | 78. | 106. |
| 23. | 51. | 79. | 107. |
| 24. | 52. | 80. | 108. |
| 25. | 53. | 81. | 109. |
| 26. | 54. | 82. | 110. |
| 27. | 55. | 83. | |
| 28. | 56. | 84. | |

Made in United States
Orlando, FL
04 November 2026

90200516R00066